Drake's
RADIO — TELEVISION
ELECTRONIC
DICTIONARY

Radio Transmission and Reception - Monochrome and Color Television - Transistors Photoelectricity - Audio Systems - High Fidelity - Electricity and Magnetism

●

Compiled by
HAROLD P. MANLY

●

DRAKE PUBLISHERS, LTD.
NEW YORK   1971

Published in 1971

Drake Publishers, Ltd., 440, Park Avenue South, New York, N.Y. 10016

SBN 87749-080-5

Printed in Taiwan
The Republic of China

PREFACE

THIS is a completely revised edition of the original Electrical and Radio Dictionary which appeared first in the days when spark transmitters still crackled and broadcast programs came in by way of a crystal detector and headphones.

During all the intervening years and editions the removal of obsolete terms has kept pace with insertion of new ones until, in the present book, emphasis is on television in color and black-and-white, on high fidelity systems, on transistors, and on the more recent aspects of sound radio. Only a few of the more ancient definitions have been retained because of their early importance and continuing historical interest.

All words, terms and abbreviations are arranged in one continuous alphabetic order. Terms consisting of more than one word follow the first word. Hyphenated words are treated as single words, while abbreviations are placed as though their letters formed a word, all being inserted in their regular order in accordance with this rule.

THE PUBLISHERS

Definitions of latest Industrial Electronics words and terms included in last pages of this book.

RADIO AND ELECTRONIC SYMBOLS

Definitions of the names of devices represented by the symbols are given in regular alphabetic order in the pages following.

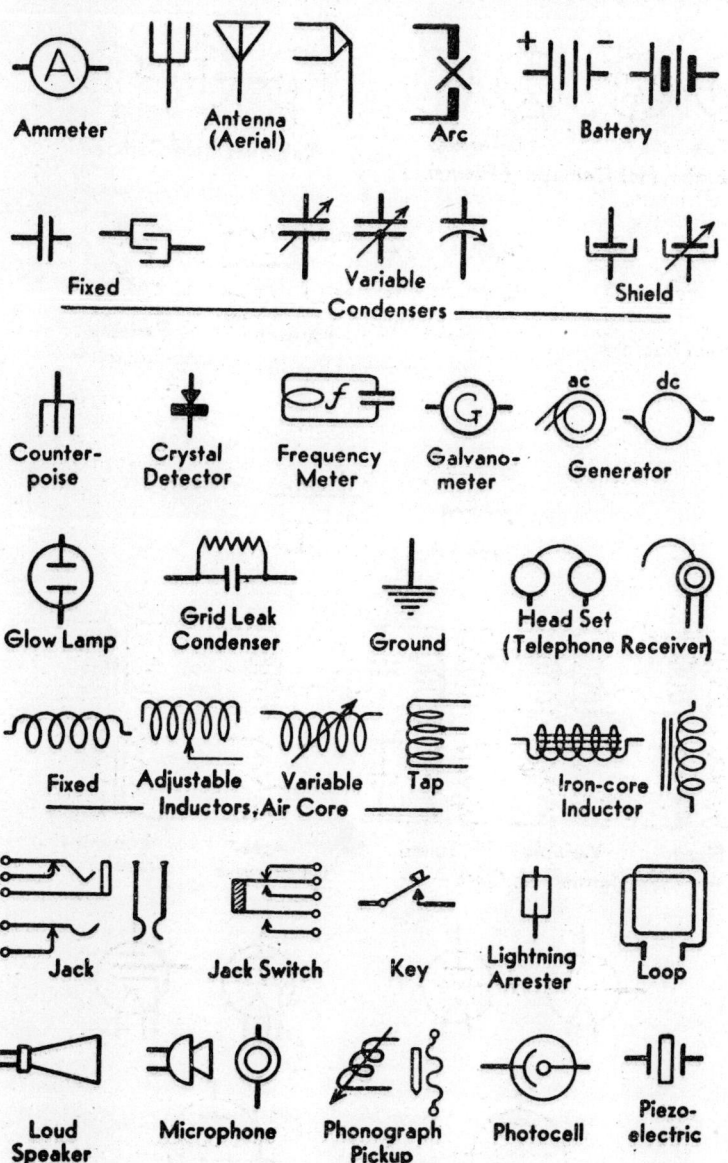

RADIO AND ELECTRONIC SYMBOLS

Where several symbols are shown as representing the same device, the symbol placed at the left or the top is preferred for use.

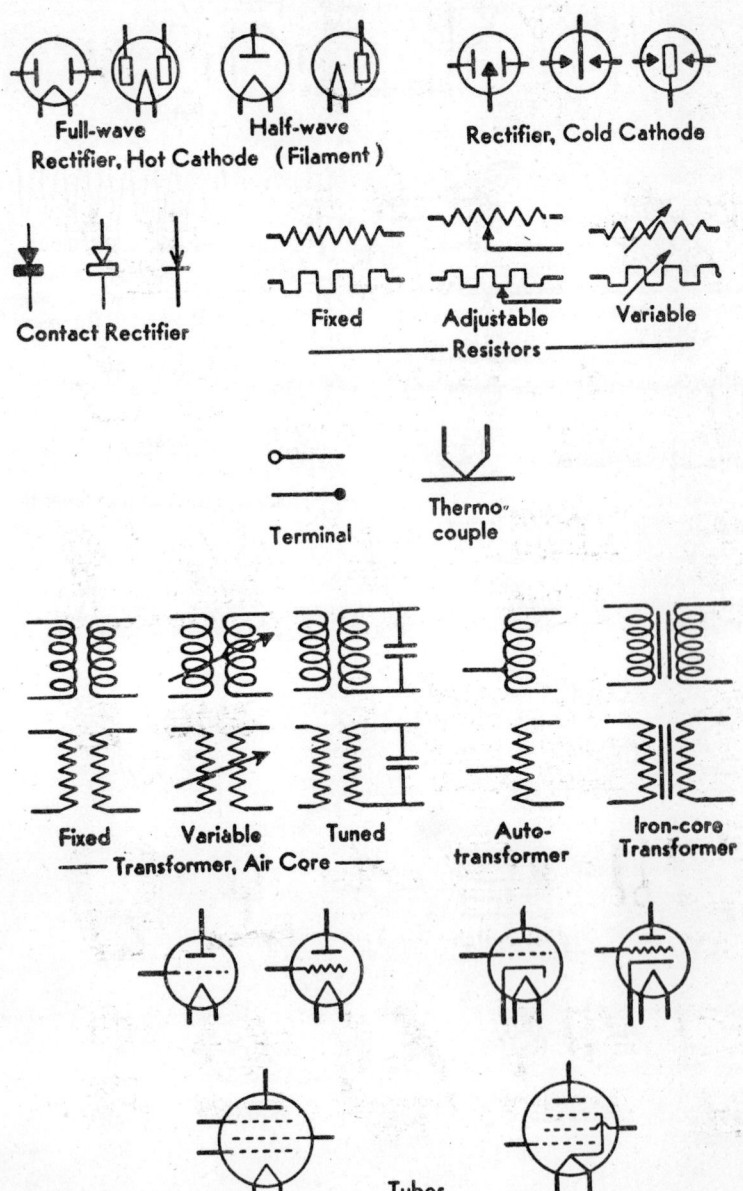

RADIO AND ELECTRONIC SYMBOLS

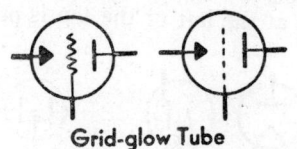

Grid-glow Tube

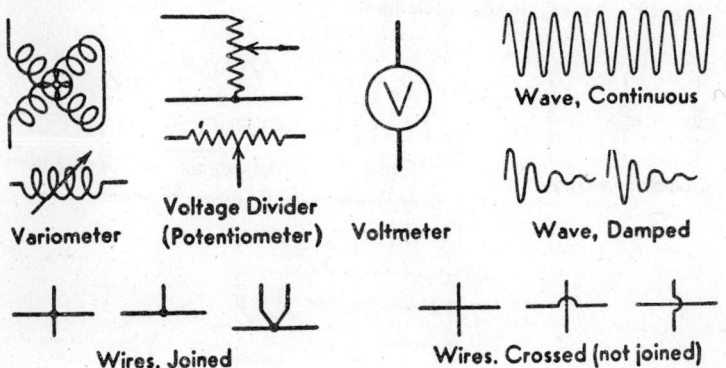

Variometer   Voltage Divider (Potentiometer)   Voltmeter   Wave, Continuous   Wave, Damped   Wires, Joined   Wires, Crossed (not joined)

DRAKE'S
RADIO-TELEVISION-ELECTRONIC DICTIONARY

A

A.—A symbol for area.
a.—A symbol for *ampere*.
ab-.—A prefix used in names of *electromagnetic units*.
A-battery.—A battery which furnishes current for the filaments of tubes.
aberration.—Failure to bring all light rays to a focus. *Chromatic aberration* or *spherical aberration*.
abscissa.—A distance measured horizontally across a graph to locate a point on a curve. See *coordinates*.
absolute efficiency.—The ratio of the output of an actual device to the output of a device similar but having no losses, both actuated by the same input.
absolute temperature.—Temperature measured from *absolute zero*, the units being centigrade degrees. The absolute temperature of a substance is equal to its centigrade temperature plus 273.1.
absolute unit.—A unit of measurement based on unchanging physical properties such as length, time, and mass. A unit in the *centimeter-gram-second* system.
absolute zero.—The temperature corresponding to the complete absence of heat; equal to 459.6 Fahrenheit degrees below zero or to 273.1 centigrade degrees below zero.
absorbed wave.—That portion of a wave or its energy which is dissipated in a medium against which it strikes.
absorbing condenser.—A condenser exhibiting *dielectric absorption*.
absorption coefficient.—The percentage of the total initial energy which is absorbed from a wave of any kind and dissipated (as heat) in a medium against which the wave strikes or through which it passes.
absorption current.—The current which flows into a condenser following its initial charge, this current being due to a gradual penetration of the electric strain into the dielectric. Also the current which flows out of a condenser following its initial discharge.

absorption marker.—A sharp dip produced on a frequency response curve by absorption of energy into a connected circuit tuned to the frequency at which the dip appears. The absorbing circuit usually is in a marker generator.

absorption unit for sound.—That dissipation for sound energy which is provided by a clear opening one foot square; energy passing through such a space being assumed as completely dissipated.

absorption wave trap.—A *wave trap* consisting of a resonant circuit inductively coupled to an antenna circuit. Power at the undesired frequency is absorbed by the trap circuit.

A.C., a.c., or a-c.—Abbreviations for *alternating current*.

acceptor.—An impurity intentionally added in minute quantity to germanium in forming a *p*-type transistor crystal. Acceptor atoms have fewer electrons which may be shared than have germanium atoms, and thus introduce positive charges which are called holes.

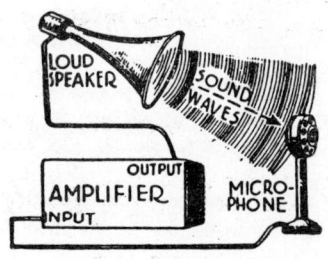
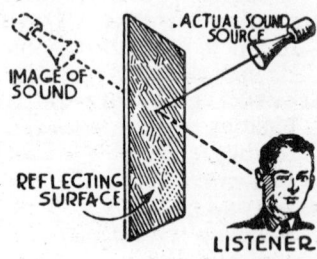

Acoustic Feedback   **Acoustic Image**

accompanying sound.—The sound signal that is a portion of a received television program. The sound carrier in the same channel with received video or picture signals.

acetate tape.—A sound recording tape having a smooth transparent plastic backing on which is the oxide coating to be magnetized.

achromatic.—Without color. A name sometimes applied to black-and-white television as distinguished from color television.

acoustic.—Pertaining to sound and hearing.

acoustic feedback.—A transfer of sound waves or energy from a loud speaker to a microphone in the same amplifying system. Any transfer of sound wave energy from a sound producing member to another member preceding it in the same system. *See illustration.*

acoustic filter.—An arrangement of sound chambers and passages which allow comparatively free passage of sounds of certain frequencies while impeding or preventing passage of other sound frequencies.

acoustic frequency.—An *audio frequency*.

ACOUSTIC IMAGE

acoustic image.—An imaginary source of sound assumed to exist at the same distance behind a reflecting surface as the distance of an actual sound source from the front of that surface, the imaginary source having the same frequency and phase as the actual sound source. *See illustration.*

acoustic impedance.—The opposition of a sound conveying medium (such as air) to motion of a sound radiating surface. Equal to the pressure developed on a *unit surface* divided by the product of its velocity and area. One unit of acoustic impedance exists when a pressure of one *bar* produces a volume velocity of one cubic centimeter per second. A similar unit is used for *acoustic resistance* and *acoustic reactance.*

acoustic labyrinth.—A speaker enclosure having partitions and passages that lengthen the path traveled by sound waves before they emerge into surrounding air.

acoustic radiator.—The portion of a loud speaker at which sound waves originate.

acoustic reactance.—That portion of the *acoustic impedance* which is due to the effective mass of the sound radiating medium.

acoustic resistance.—That portion of the *acoustic impedance* which results in dissipation of energy. The real component of the acoustic impedance.

acoustic resonance.—An increase of sound intensity due to combination of *reflected waves* and *direct waves* which are in phase, or due to vibration of air columns or solid bodies at the sound frequency.

acoustic resonator.—An enclosure which increases the intensity of sounds at frequencies for which the enclosed air is set into natural vibration.

acoustic wave.—A sound wave in a gas, a liquid or a solid. A moving variation in the density, pressure or velocity of a medium.

acoustics.—1—The science of sound and its effects. 2—The properties of a room or an enclosure as they affect the behavior of sounds within the room.

active circuit.—The portion of a circuit which furnishes resistance, inductance or capacity essential to functioning.

active current or component.—The portion of the current which is *in phase* with the applied voltage, or the voltage which is in phase with the current it produces in an alternating current circuit. The part of the current or voltage which produces power in *watts*. Compare *reactive current or component*.

active lines.—In a television picture tube or camera tube, the periods during which the electron beam is scanning the lights and shades of the picture or is reproducing them.

active transducer.—A *transducer* supplying power from itself to a second system, this power being controlled from that in the first system. An amplifier using vacuum tubes is such a device.

activity coefficient.—The *space factor*.

ACYCLIC

acyclic.—Not following a cycle; *aperiodic*.

adapter.—Any device allowing the use of a part in a position or function for which that part is not originally intended.

adder.—In color television receivers, an amplifier circuit in whose input are combined definite portions of luminance and chrominance signals for producing primary color signals delivered to picture tube grid-cathode circuits.

additive process.—A method of producing colored motion pictures in which shades and tones of different colors are formed by superimposing two or more *primary colors*.

adjacent sound.—The television sound carrier signal in the channel next higher in number than the channel to which the receiver is tuned.

adjacent video.—The television video carrier in the channel next lower in number than the channel to which the receiver is tuned.

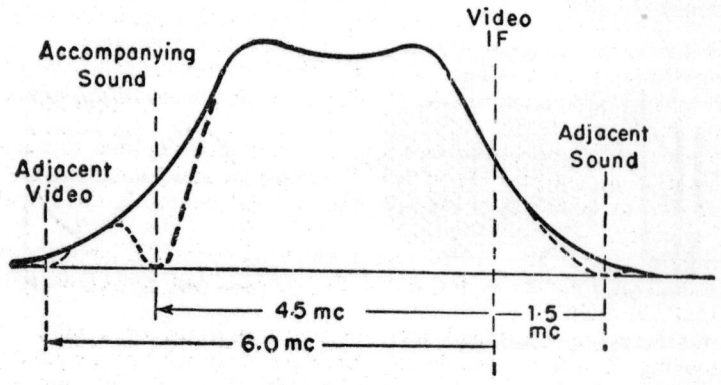

Adjacent Sound and Video

adjustable resistor.—A resistor of which the resistance is adjusted manually.

Admiralty unit.—A British unit of *capacity*, equal to 0.0011 microfarad.

admittance.—A measure of the ease with which an alternating current flows in a circuit. The reciprocal of the *impedance*. Measured in *mhos*. The usual symbol is Y or y.

aerial.—1—The overhead conductors of a *capacity antenna* system. 2—Sometimes used to mean the entire antenna. See also *antenna*.

A.F., a.f., or a-f.—Abbreviations for *audio frequency*.

afc or a.f.c.—Abbreviation for *automatic frequency control*.

afterimage.—The impression of sight which remains in the eye after the image is no longer visible. The result of *persistence of vision*.

agc or a.g.c.—Abbreviation for *automatic gain control*.

aging.—1—An increase of *hysteresis* and lessening of *permeability* in iron cores. 2—In *reactivation of filaments* that part of the process in which a thoriated filament is operated at a moderately high voltage to form a fresh layer of thorium. 3—In tube manufacture, a period of operation under conditions approximating normal service to bring the electrodes into normal condition or to determine operating characteristics.

A. I. E. E.—Abbreviation for *American Institute of Electrical Engineers.*

air column.—The body of air contained within a horn or other sound chamber.

air condenser.—A condenser having air at atmospheric pressure for its dielectric.

air-core.—Descriptive of an inductance coil or transformer having no iron in the magnetic circuit. A construction used in high frequency circuits.

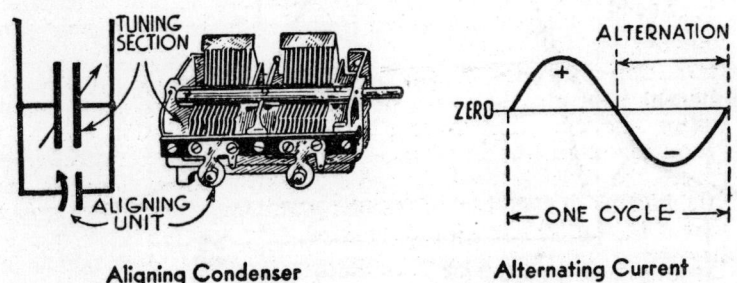

Aligning Condenser   **Alternating Current**

air-dielectric.—Descriptive of a capacitor, usually adjustable, in which air forms the dielectric between plates.

air gap.—A small space filled with non-magnetic material, forming a portion of a magnetic circuit otherwise completed through iron. The gap reduces tendency toward *saturation* and makes the magnetic flux more nearly proportional to the magnetizing force.

aligning condenser.—An adjustable condenser of small capacity connected in parallel with one unit of a *gang condenser* so that the total capacity of the section may be adjusted to allow simultaneous tuning with other sections. *See illustration.*

alignment.—Adjustment of capacitors, inductors, and resistors in radio and television circuits for allowing resonance at certain frequencies, for providing desired pass bands, for rejecting various frequencies or signals, and, in general, for obtaining normal operation or performance.

alkali earth.—One of a group of metals which includes *barium, calcium, magnesium and strontium.* They are *photo-emmisive* when acted upon by certain wavelengths of radiant energy.

alkali metal.—A metal of the alkali group, capable of reacting with an acid to form a salt of the metal. *Potassium, sodium, lithium, rubidium, caesium.* These metals are *photo-emissive.*

allocation.—The frequency or band of frequencies at or within which a radio or television transmitter is permitted to radiate carrier signals.

alpha.—A symbol for current amplification factor between emitter and collector of a transistor. The ratio of change in collector current to change of emitter current when collector voltage is held constant. In point-contact transistors alpha is greater than 1.0, while in junction transistors it is less than, but approaches 1.0.

alpha rays.—*Radioactive rays* consisting of positively charged helium atoms; rays having comparatively little penetration through opaque objects but a decided ionizing effect on gases.

alternating component.—In a *pulsating current* or voltage or in a regularly varying *direct current* or voltage, the alternating quantity represented by rise and fall above and below the average value considered as zero for the alternations. The alternating current or voltage which would remain with the *direct component* filtered out.

alternating current.—A current which reverses its direction at regular intervals, the waves of opposite polarity having the same shape and size. The average value of an alternating current is zero. One complete set of values in both directions constitutes one cycle, the time required for one cycle is the current's period and the number of cycles in one second is the frequency. The symbols are *A.C., a.c.,* or *a-c. See illustration.*

alternating current power supply.—A *power unit* taking its energy from an alternating current line.

alternating current resistance.—*Impedance* at a certain frequency.

alternating quantity.—An electric current or voltage, a magnetic force or flux, or other values which regularly and periodically reverse their direction.

alternation.—One-half of an alternating current *cycle,* during which current or voltage changes from zero to maximum in one direction and then back to zero.

aluminized screen.—A television picture tube viewing screen which is coated, on the back or inner side of the phosphor, with an exceedingly thin layer of metallic aluminum. Phosphor illumination which otherwise would be lost within the tube is reflected outward with resulting increase of brightness and range of contrast in pictures.

aluminum rectifier.—An *electrolytic rectifier* employing aluminum as one of its elements.

amalgam.—A combination of mercury with another metal.

amateur.—The operator of a short wave or high frequency transmitter which is privately owned and is not used for commercial work.

ambient temperature.—Temperature of air or liquid surrounding any electrical part or device. The term usually refers to the effect of such temperature in aiding or retarding removal of heat by radiation and convection from the part or device in question.

American Morse code.—A system of telegraphic signals used chiefly in wire telegraphy and differing from the continental code in some of the letters and in the method of spacing.

American wire gage.—The gage or standard of measurement for the diameter of wire conductors of copper, brass, bronze, etc. Brown & Sharpe wire gage.

ammeter.—An instrument for measuring current flow in *amperes*.

ammeter shunt.—A resistance which remains nearly constant under all conditions, placed between the terminals of an ammeter to carry the bulk of current being measured. *See illustration.*

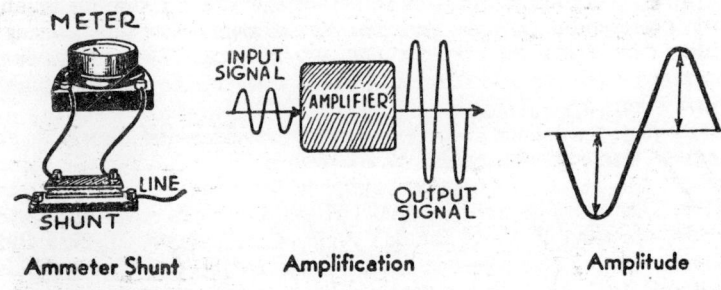

Ammeter Shunt   Amplification   Amplitude

amperage.—The current in *amperes*.

ampere.—The practical unit of rate of flow of electric current. The current flowing through one ohm resistance with one volt pressure. A flow of one coulomb per second. The *International ampere*. The symbol is a.

ampere-hour.—The quantity of electricity passing with a flow of one ampere continued for one hour. Equal to 3600 *coulombs*. The symbol for quantity of electricity in ampere-hours or coulombs is Q.

ampere-turn.—A unit of *magnetomotive force*. The force produced by one ampere of current flowing in one complete turn of a conductor. The number of ampere-turns is equal to the product of the number of turns in a winding and the number of amperes flowing in it. One ampere-turn is equal approximately to 1.257 *gilberts*.

amplification.—An increase in the voltage, current or power of a signal. Amplification occurs in vacuum tubes and transformers. *See illustration.*

amplification coefficient, constant or factor.—A measure of the effect of grid voltage changes as compared with the effect of plate voltage changes in producing a given change of plate current in a vacuum tube. The ratio of a plate voltage change to a grid voltage change, both of which result in equal variation of plate current with all other factors remaining the same. The ratio of the alternating voltage appearing in the plate circuit to the alternating voltage applied to the grid circuit of a tube with which the plate circuit load is an infinite impedance. The symbol is the Greek letter *mu* (μ).

amplified automatic gain control.—Automatic gain control with which voltage that varies with signal strength, and varies grid bias on controlled tubes, is strengthened by a tube used expressly for this purpose.

amplifier.—The means by which signals are strengthened in power, voltage or current; the input power controlling a larger power supplied locally from the amplifier circuits.

amplify.—To increase the strength of received or transmitted signals.

amplifying transformer.—A transformer having a step-up voltage ratio, generally an *audio frequency transformer*.

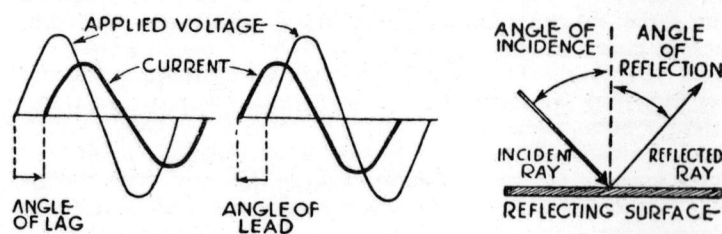

Angle of Lag and Lead   Angle of Light

amplifying tube.—A vacuum tube used for the purpose of increasing the voltage, current or power applied to its grid circuit.

amplitude.—The greatest value of an alternating quantity in one direction or in one polarity. A *peak value*. See illustration.

amplitude distortion.—A type of *distortion* occurring when high voltages in the applied signals are multiplied by a different factor than the low voltages, the ratio of gain or of loss changing with changes in signal *amplitude*.

amplitude modulation.—Varying the strength or amplitude of a carrier wave at frequency of a transmitted signal. Signal envelope waveforms correspond to the modulating signal. See *modulated wave*.

angle.—1—of lag and of lead: The number of *electrical degrees* by which the current in an alternating current circuit reaches its peak and zero values after (angle of lag) or before (angle of lead) the applied voltage reaches the corresponding peak and zero values in the same circuit. See *inductive load* and *capacitive load*. See illus-

tration. 2—*of incidence:* The angle between a line perpendicular to a reflecting surface and a light ray striking that surface. 3—*of reflection:* The angle between a line perpendicular to a reflecting surface and a light ray reflected from that surface. *See illustration.*

angle of deflection.—The *deflection angle.*

Angstrom unit.—A unit of *wavelength,* equal to the one hundred-millionth part of a *centimeter* or to one ten-thousandth of a *micron.*

angular velocity.—In an alternating current circuit, the frequency multiplied by 2π, or by approximately 6.2832. Measured in *radians* per second. The symbol is the Greek letter omega (ω).

anion.—A *negative ion;* an ion which moves toward an anode.

anode.—The electrode or element from which electrons leave an electrical device. The plate of a vacuum or any electronic tube. The negative terminal of a battery or other source.

antenna.—Conductors arranged to absorb energy from a radio wave or to radiate energy in the form of radio waves.

antenna capacity.—The *electrostatic capacity* of the antenna circuit.

antenna circuit.—The entire path traversed by currents produced in an aerial, its ground and any capacities, inductances and resistances in circuit with these two.

antenna coupling.—The coupling through which there is transfer of energy from an antenna circuit to a receiver, or from a transmitter to its antenna circuit.

antenna current.—The current flowing in an *antenna circuit.*

antenna gain.—The ratio of signal voltage delivered from the antenna considered to voltage delivered from a standard or reference antenna when both antennas are subjected to the same space signal and when both deliver their signal outputs to the same or equivalent receiving systems or measuring devices. The reference antenna usually is a dipole of electrical length corresponding to wavelength of each signal frequency at which comparisons are made.

antenna impedance.—Impedance of an antenna at its resonant frequency, which is the frequency at which capacitive and inductive reactances of the antenna become equal and cancel to leave high-frequency resistance as the only factor in impedance.

antenna effect.—The ability of any conductor insulated from the earth to act as one plate of a *capacity antenna.* Especially, such an effect as observed in a *loop antenna.*

antenna lead-in.—The conductor connecting an aerial wire to a receiver. Compare *down lead. See illustration.*

antenna reflector.—That part of a *beam antenna* which reflects or reverses the direction of the radiation.

antenna resistance.—The total *effective resistance* of an antenna circuit. The resistance found by dividing the average power in an antenna circuit by the square of the maximum effective current in the antenna. Antenna resistance includes the radiation resistance, the ground resistance, the losses due to corona, eddy currents, dielectric effects and insulation leakage, and also the ohmic resistances of conductors.

antenna switch.—A switch for disconnecting the aerial wire from a receiver. The same switch may ground the antenna. *See illustration.*

antenna tower.—The supports for the electrical portions of an aerial.

antenna tuning condenser or tuning inductance.—A variable condenser or variable inductance used in an antenna circuit to produce *resonance* at certain desired frequencies. *See illustration.*

antenna wavelength.—The wavelength corresponding to the *natural frequency* of the antenna circuit.

anti-capacity switch.—Any form of switch in which the metallic current carrying parts are well separated to reduce the *electro-*

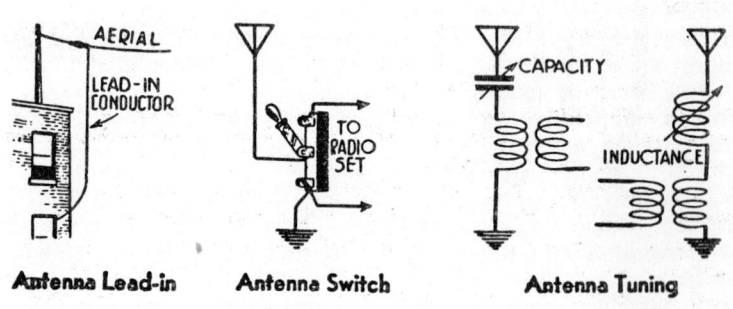

Antenna Lead-in   Antenna Switch   Antenna Tuning

static capacity between them. Usually a cam switch.

antinode.—The points in a series of waves where the greatest amplitude or field strength exists. Points midway between the nodes. See *node*.

antiphase.—*Opposite phase.*

anti-regeneration.—The process of applying energy from a tube's plate circuit to its grid circuit with the voltages in such phase relation that the grid voltage changes are reduced in strength.

anti-resonance.—*Parallel resonance.*

aperiodic.—Having no period, no natural frequency. Not *resonant* at any one frequency.

aperiodic circuit.—A circuit which is not tuned or *resonant*. A circuit in which oscillations are not maintained, the resistance serving to damp out oscillatory effects. A circuit is aperiodic when the resistance squared is greater than four times the inductance in henries divided by the capacity in farads.

aperiodic instrument.—A *deadbeat instrument.*

apparent inductance.—The combined effect of a coil's *true inductance* and its *distributed capacity* in determining the resonance frequency of a circuit containing the coil. That value of inductance which would have the same effect on tuning as is actually had by a coil's real inductance and its distributed capacity together. Compare *true inductance*.

aperture mask.—In a three-gun color television picture tube, just behind the phosphor plate or screen, a thin opaque plate with a number of openings equal to the number of groups of phosphor dots on the phosphor screen. Each opening is large enough to pass the three beams at their point of convergence. *See illustration.*

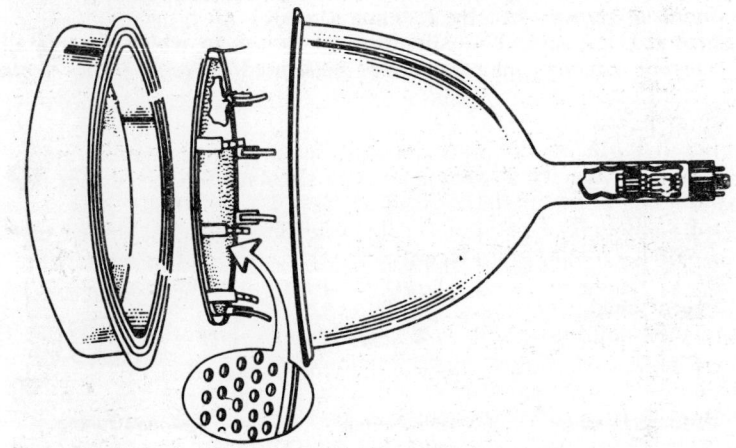

Aperture Mask

apparent power.—The *volt-amperes* in an alternating circuit.

apparent resistance.—The *impedance* of an alternating current circuit.

applied voltage.—The *impressed voltage*.

arc.—A luminous discharge or continuous flow of current across a space between electrodes, the conduction being due chiefly to ionization of gases in the space. A very intense or concentrated *glow discharge*.

arc over.—An arc or spark that passes from any point on a high-voltage conductor to any other conductor at lower potential. A term often used with reference to television picture tube high-voltage circuits.

armored cable.—An insulated conductor protected with a twisted or braided covering of metal.

array.—All the active elements in an antenna system, including dipoles which feed signals to a receiver, also any associated reflectors and directors. An array may include two or more bays.

argon.—One of the inert gases existing naturally in the atmosphere.

argon rectifier.—A hot cathode type of *gaseous conduction rectifier* using argon gas in the bulb.

arithmetic mean.—One-half the sum of two values; the average.

armature.—1—The part of a generator or motor carrying the conductors in which induction results in electromotive force or in power. 2—A piece of iron or steel forming part of a magnetic circuit, being either fixed in position or moved under the influence of changing magnetic strength.

articulation.—The percentage of detached speech syllables which may be correctly understood by a listener when the only source of distortion between the original sound source and the listener is the device of which the articulation is being measured.

artificial antenna.—A circuit having values of resistance, inductance and capacity equivalent to those of a transmitting or a receiving antenna, but from which energy is dissipated in the form of heat rather than in radio waves. Used for testing purposes. See *standard antenna*.

artificial cable or line.—A network containing resistance, inductance and capacity so arranged as to simulate the effect of a real transmission line in electrical effect.

aspect ratio.—The ratio of width to height of images formed in a television camera, and, accordingly, this ratio in transmitted pictures. The standard aspect ratio is 4-to-3, four units of width to three of height.

associated sound.—Same as *accompanying sound*.

astatic.—Having little or no magnetic property.

astatic coil.—An *inductance coil* having a very limited external field due to its construction with two cylinderical windings, one within the other, so connected that the two polarities oppose. The outer winding acts as an effective shield for the inner one. See *illustration, page following*.

astatic galvanometer.—A galvanometer with two sets of magnets which make it independent of outside magnetic influences.

astigmatism.—A change of shape of the focused spot from an electron beam as the spot moves to various positions on the viewing screen of a television picture tube or a cathode-ray tube. The spot becomes elongated vertically or horizontally, rather than remaining circular, at various areas of the screen.

asymmetric.—Not *symmetrical*. *Unilateral*.

asymmetric conductor.—A conductor allowing more current to flow in one direction than in the other.

asymmetrical antenna.—An antenna which radiates or receives more effectively in some directions than in others due to its form of construction.

atmospheric absorption.—Dissipation of energy from a radio wave moving through space, the absorption being due to atmospheric

conductivity and to the consequent *reflection* and *refraction* of the waves.

atmospherics.—Radio waves produced by electrical effects in the atmosphere; a form of *static* interference.

atom.—The smallest particle of an elementary substance, a particle which is not further divisible by chemical means. Compare *molecule*. An atom consists of a positive nucleus around which rotate one or more negative electrons.

atomic weight.—The weight of an atom of a substance in comparison with the weight of an atom of hydrogen.

attenuation.—1—A lessening of intensity or power. Any reduction in voltage, current or power between different circuits or between different parts of one circuit. 2—Radio attenuation is the decrease in *amplitude* of the electric and magnetic forces in a radio wave with increase of distance from the transmitter.

attenuation box.—An *attenuation network* provided with switching means for altering the power loss. *See illustration.*

attenuation coefficient, constant or factor.—1—The rate at which a wave diminishes in energy with distance from the transmitter or source. 2—The relation between the current at a source and the current received from a uniform transmission line. The attenuation occurring in one section of a line of recurrent structures. 3—The number which, multiplied by the distance of transmission, gives the natural logarithm of the ratio of the amplitude of the forces at that distance to their amplitude at the transmitter

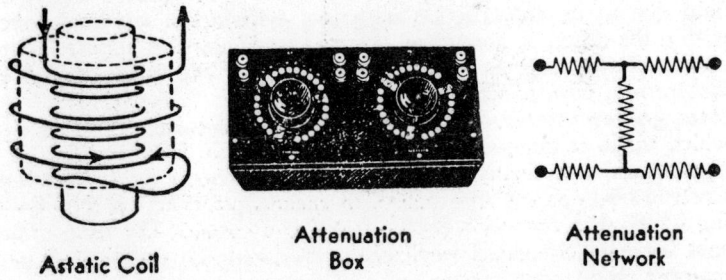

Astatic Coil   Attenuation Box   Attenuation Network

attenuation equalizer.—A device which increases the *transmission loss* at certain frequencies so that the loss for all transmitted frequencies within a certain range is made substantially the same. Frequencies originally over-emphasized are reduced in amplitude to equal all other frequencies. A system of inductances, capacities and resistances which attenuates certain frequencies.

attenuation network.—An arrangement of resistors used to introduce a known amount of power loss in a circuit while maintaining desired impedance relations. *See illustration.*

attenuation pad.—An *attenuation network*.

attenuator.—A device for lessening the amplitude of a signal current while introducing minimum distortion.

attraction.—*Electric attraction*.

audibility.—The strength of a signal or sound as measured in units based on the response of the ear. The ratio of the actual signal strength to the strength required for a signal which can barely be heard.

audibility factor.—The ratio between the resistances employed in a test with an *audibility meter*. A measure of audibility which is made by varying a resistance connected in parallel with a telephone receiver.

audibility meter.—A variable resistance connected across a telephone receiver and adjusted so that signals from a measured source are just audible. The resistance required allows comparison of strength between a measured source and a reference source as a standard. *See illustration*.

audible.—Capable of being heard; perceptible to the ear.

audible spectrum.—Sound waves within the limits of human hearing; the *audio frequencies*.

audio frequency.—A current, voltage, sound or other wave motion having a frequency which affects the ear as sound, approximately

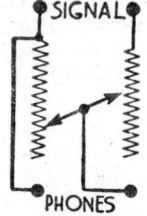
Audibility Meter

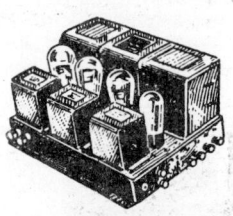
Audio Frequency Amplifier

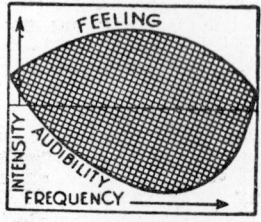
Auditory Sensation Area

between 20 and 20,000 cycles per second. Abbreviations are *A.F.*, *a.f.*, and *a-f*.

audio frequency amplification.—An increase in voltage, current or power of a signal at *audio frequency*.

audio frequency amplifier.—Apparatus containing vacuum tubes and used for increasing the voltage and power of an *audio frequency* signal. *See illustration*.

audio frequency choke.—An inductance coil used to impede the flow of *audio frequency* currents through a circuit.

audio frequency oscillator.—Apparatus for the production of currents at audio frequencies. A *beat frequency oscillator*, a **magneto-**

striction oscillator, a *tuning fork oscillator* or a *vacuum tube oscillator* working at *audio frequency*.

audio frequency transformer.—A transformer designed to provide coupling between two circuits operating at *audio frequencies*.

audio howler.—An *audio frequency oscillator*.

audiometer.—An instrument for measuring the intensity or audibility of sounds.

audition.—The act of hearing or of listening to sounds.

auditory.—Pertaining to the ear and the sense of hearing.

auditory masking.—The *masking effect*.

auditory sensation area.—The sound intensities included between the *threshold sound intensities*, between the threshold of audibility and the threshold of feeling. *See illustration, page preceding.*

aural.—Pertaining to the sense of hearing.

aurora.—The aurora borealis or aurora australis; a luminous effect of electrical origin appearing in the earth's polar regions.

Austin-Cohen formula.—A formula for calculation of the current produced in a receiving antenna when the known factors include distance to transmitter, carrier frequency, transmitter power and antenna heights.

autodyne frequency meter.—A *frequency meter* producing within itself currents at radio frequency which beat with received currents, *resonance* being indicated when the two frequencies are alike and produce a *zero beat*.

autodyne reception.—Radio reception by combining a received frequency with another frequency produced in the detector circuit of the receiver to produce a beat frequency. *Heterodyne reception* in which a single vacuum tube operates both as an oscillator and as a *first detector*. Self-heterodyne reception. The principle employed in some superheterodyne receivers.

automatic focus.—Electrostatic focusing in television picture tubes without the help of any external voltage. The focusing anode is internally connected through a fixed resistor to the cathode in the electron gun.

automatic frequency control.—Means for maintaining the operating frequency of an oscillator very closely in time or synchronization with that of some external voltage. In principle, the external synchronizing frequency, or waveforms derived from it, are combined in a control tube with a waveform of actual output from the oscillator. Any difference between frequencies produces in control tube circuits a correction voltage that is applied to the oscillator input in such a manner as will change oscillation frequency until it matches synchronizing frequency. Used with television receiver sweep oscillator and sometimes with r-f oscillators in television or radio tuners.

automatic gain control.—In television receivers, means for increasing amplification in radio-frequency or intermediate-frequency amplifier stages when strength of received signals drops,

AUTOMATIC PHASE CONTROL

and of decreasing amplification when signal strength rises. Part of the amplifier i-f signal is rectified to produce a direct voltage which becomes more negative with strong signals and less negative with weak signals. This voltage alters the grid voltage of controlled tubes. Generally similar to automatic volume control for sound radio.

automatic phase control.—In color television receivers, a control for the color oscillator acting in principle like an automatic frequency control, but maintaining oscillator output voltage synchronized in phase as well as frequency with phase and frequency of the burst signal from the transmitter.

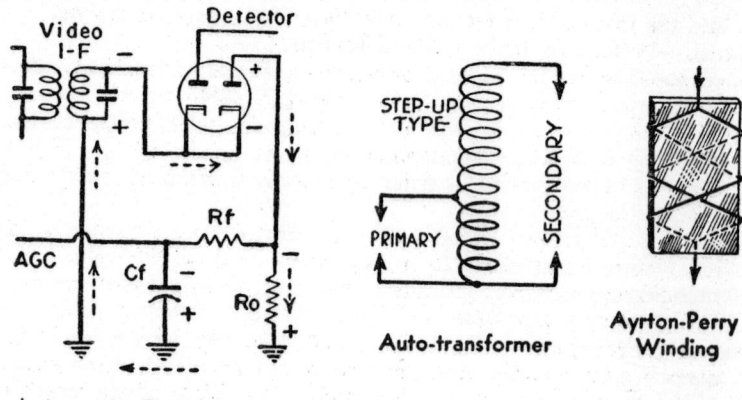

Certain stations or frequencies are tuned in by press buttons, keys or similar devices.

automatic volume control.—A method of automatically maintaining a nearly constant power output or loud speaker volume from a receiver in which the antenna input is varied by fading and similar effects. With one commonly used system a part of the signal is rectified and used to alter the control *grid bias* on radio frequency or intermediate frequency amplifying tubes. Increase of signal strength then makes the bias more negative which reduces the amplification, while a smaller antenna input produces a contrary effect, amplification being made inversely proportional to antenna input.

autoplex receiver.— super-regenerative receiver.

auto-transformer.—A transformer in which part of the primary winding is also a part of the secondary, or vice versa, the two windings being conductively connected together so that a part of the energy in the secondary comes directly from the primary. *See illustration.*

avc or a.v.c.—Abbreviation for *automatic volume control*.

average value.—The algebraic sum of two opposite effects, such as currents or voltages of opposite polarity. The difference, if any, between the values of one polarity and those of the opposite polarity. The average value of a sine wave alternation is equal to 0.637 times the maximum or peak value.

Ayrton-Perry winding.—A *non-inductive winding* using two conductors connected in parallel, one of them carrying current clockwise and the other carrying it anti-clockwise around the winding. *See illustration, page preceding.*

B

B.—A symbol for *magnetic flux density*.

b.—A symbol for *susceptance* in mhos.

back coupling.—A coupling by means of which energy from an output circuit is applied to the input circuit of the same system. *feedback*.

back current.—Current which flows through a rectifying element in a direction opposite to that in which the rectified current is passed. The current which results from those portions of alternations which are not completely suppressed by the valve action of a *rectifier*.

back e.m.f.—*Counter-electromotive force*.

back focal length.—The distance from the center of a lens to its *principal focus* on the side of the lens away from the object.

background noise.—Noises which result from irregular and slight changes taking place in the circuit elements of a transmitter. Also *tube noise, needle scratch* and similar effects.

back porch.—In a composite television signal, the portion of the pedestal in the horizontal blanking interval that follows a horizontal sync pulse.

baffle.—A partition used to prevent free movement of air between front and back surfaces of a *free radiator*. The baffle surrounds and extends outward from that part of a loud speaker which originates the sound waves so that air compressions on one side of the radiator cannot be neutralized by the rarefactions simultaneously produced on the other side. *See illustration*.

bakelite.—An insulating material composed chiefly of *phenolic compound*.

balanced amplifier.—An amplifier using *push-pull amplification*.

balanced H-section attenuator.—An *attenuation network* maintaining constant impedance relations in both directions and allowing balance of both sides of the line to ground. *See illustration*.

balanced modulator.—A modulation system used with *side band transmission*. Two tubes are connected as in a push-pull circuit, the carrier being eliminated in their output transformer.

balanced transmission line.—Two side by side ungrounded conductors in which, when the line extends vertically, interference fields or impulses tend to induce currents the same direction in the line conductors, but of opposite polarities at the receiver signal input terminals. Interference currents thus tend to balance or cancel.

BALANCING

balancing.—The process of adjusting an external capacity so that it allows a *feedback* of voltage which is opposite in phase and equal in effect to the feedback through the plate-grid capacitance of an amplifying tube.

ballast.—A resistor, usually connected in series with series heaters in tubes, to limit the surge of current which otherwise would occur when power is turned on, and while heater resistance still is low due to the temperature. The ballast may be a negative temperature coefficient resistor, or may be iron wire that quickly attains red heat and high resistance, which drops when current drops to normal.

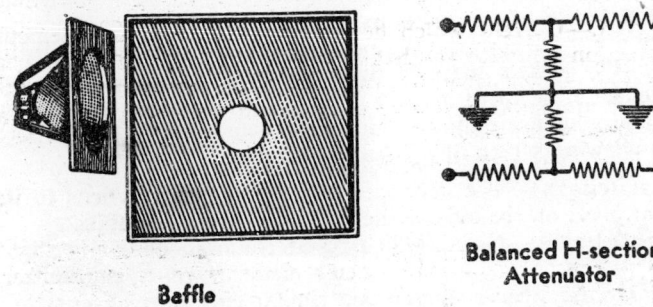

Baffle

Balanced H-section Attenuator

ballast tube.—A tube containing a resistance element which maintains a fairly constant current in its circuit when changes occur in the applied voltage.

ballata.—A natural vegetable gum similar to *gutta-percha*.

ballistic galvanometer.—An instrument for measuring quantity of electricity in *coulombs*. The movement is undamped, and a single swing of the pointer is proportional in extent to the amount of electricity discharged from a condenser through the movement.

band exclusion filter.—An electric filter which greatly attenuates currents between two limiting or *cutoff frequencies* flowing in the circuit of which the filter is a part. *See illustration.*

bandpass amplifier.—In a color television receiver, the section whose circuits are tuned to amplify and pass only the frequencies of the chrominance signal, not those of the entire luminance signal or video signal.

band pass filter.—An electric filter which allows comparatively free passage of currents at frequencies within certain upper and lower limits while greatly attenuating currents at all frequencies outside these limiting or *cut-off frequencies*. *See illustration.*

bandspread.—A tuning control whose entire range includes only a small portion of the frequency band in which a receiver may operate, thus making it possible to tune accurately and conveniently within the spread range.

bandwidth.—The range of frequencies within which gain of an amplifier system remains above some certain fraction of its maximum value. The range of frequencies below the low and high limits at which amplification or gain drops to a specified fraction of maximum, the fraction commonly being 0.707.

banked winding.—A coil winding in which successive turns are laid up into two or more layers, the whole winding proceeding in this manner from one end to the other of the form without returning at any point. Employed to reduce the *distributed capacity* of a winding requiring many turns. *See illustration.*

bar.—A pressure of one dyne per square centimeter.

bar generator.—A color television service instrument furnishing voltages whose phase relations with respect to a reference frequency are those corresponding to certain phases of a chrominance signal. The phases may be those for color primaries and their complements, or there may be a continual variation through the entire chrominance range. Color bars or bands are made to appear on the viewing screen of a color picture tube.

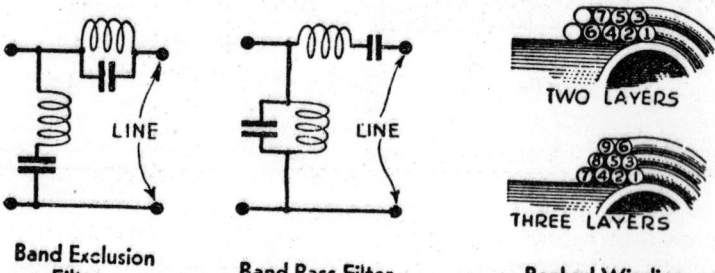

Band Exclusion Filter Band Pass Filter Banked Winding

barium.—An alkaline earth of which compounds are used in cathodes of photocells having maximum response to yellow-green light.

Barkhausen effect.—Slight fluctuations in the rate at which a material becomes magnetized, the magnetization not proceeding at a perfectly uniform rate.

Barkhausen oscillation.—A variety of parasitic oscillation sometimes occurring in television horizontal output amplifier tubes and circuits, causing a narrow, dark, vertical line near the left-hand side of pictures and rasters.

base.—The crystal element in a point-contact transistor, or the central or middle element in a junction transistor. Connection to the base for external circuits is a metallic conductor attached to the base element.

bass boost.—In an audio-frequency amplifier, such values or combinations of capacitance, resistance, and sometimes inductance.

with or without resonance effects, as allow greater amplification at low frequencies than at middle or high audio frequencies. Also a means for attenuating middle and high frequencies for apparent emphasis of lower frequencies.

bass reflex.—A speaker cabinet designed to bring into phase the front and back sound waves from a speaker as they pass into surrounding air space. *See illustration.*

battery.—One or more electrically connected *primary cells* or *storage cells* in which chemical energy is changed into electromotive force.

bay.—A group of antenna conductors including one element from which signals are delivered to a receiver line, together with any reflectors and directors associated with that element. One of several similar groups in an array.

B-battery.—A battery which furnishes current for the *plate circuits* of vacuum tubes.

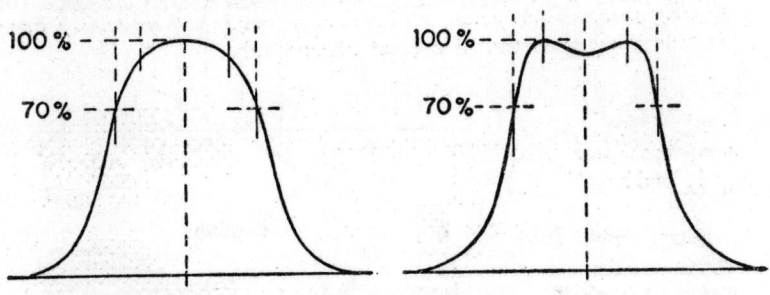

Band Width

beam antenna.—An antenna from which radiation is confined to a small angle in one direction only.

beam bender.—An *ion trap magnet.*

beam cutoff.—In a television picture tube or cathode-ray tube, prevention of an electron beam by a control grid so negative in relation to existing ultor voltage as to prevent electrons passing from electron gun to viewing screen.

beam power tube.—A tube in which electron streams from cathode to plate are concentrated by means of confining electrodes. In four-element beam power tubes secondary emission from the plate is suppressed by electrons whose velocity is retarded in a region between screen and plate to give the effect of a negative space charge in this region. In five-element beam power tubes. there is a suppressor grid to retard secondary emission.

beam switching.—Causing the electron beam of a single-gun color television picture tube to move back and forth over adjacent phosphor strips for the three primary colors while the beam is deflected horizontally.

beam transmission.—Radiation of radio waves in a direction limited to a small angle instead of almost equally in all directions. The result of wave *reflection* from a system of conductors back of the radiating antenna.

beat.—One cycle of a wave produced by *beating*.

beat frequency.—A frequency equal to the difference between two other frequencies *beating* together.

beat frequency oscillator.—Two oscillators whose output frequencies are combined, by beating action in a mixer, to produce a desired lower frequency equal to the difference between oscillator frequencies.

beat frequency receiver.—A *superheterodyne receiver*.

beat interference.—*Beating* together of two modulated carrier

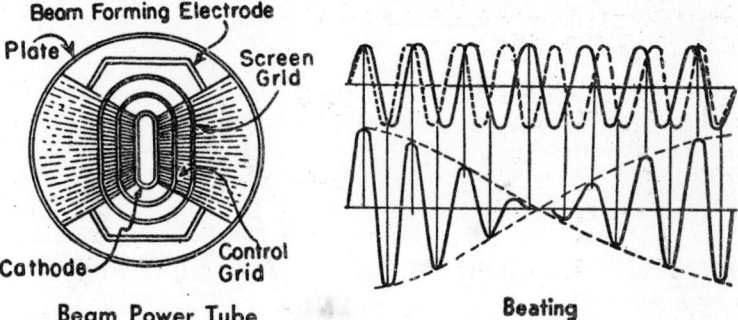

Beam Power Tube   Beating

waves to produce a new frequency in the radio frequency amplifier of a receiver, this beat frequency carrying one or both the original modulations and being detected in the usual manner. See also *cross modulation*.

beat note.—An audible frequency produced by the *beating* of two higher frequencies.

beat reception.—1—*Heterodyne reception.* 2—*Zero beat reception.*

beating.—The action by which two or more waves of different frequencies acting together produce an additional frequency equal to the difference between two of the original frequencies. The instant at which peak values of like polarity occur simultaneously in the original frequencies is the instant at which there is a peak amplitude in the beat frequency. *See illustration.*

Becquerel effect.—The *photo-voltaic* action.

beeswax.—An insulating wax having fairly low dielectric loss. Dielectric constant about 3.0.

bel.—A transmission unit in the decimal system. The commonly used unit is a *decibel*, equal to one-tenth of a bel.

bending.—In television pictures, inclination to either right or left, at and near the top of the viewing screen, of lines which should be vertically straight.

bent gun ion trap.—An ion trap in which ions and electrons are initially directed toward the inner wall of the electron gun by inclining, with respect to the axis of the picture tube, a portion of the gun toward the tube base. Electrons are turned to their correct path by the field of a single external magnet, while ions are trapped in the gun.

beta.—A symbol for current amplification factor between base and collector of a transistor.

beta rays.—*Radioactive rays* consisting of negatively charged particles, or of electrons. *Cathode rays.*

B-H curve.—A *magnetization curve.*

bias.—A difference in potential between electrodes of a vacuum tube. See *grid bias.*

bias rectifier.—A rectifier, either contact or tube, that produces direct voltage for grid biasing when furnished with alternating voltage such as that from a tube heater circuit.

bifilar transformer.—A high-frequency coupling transformer having insulated wires for primary and secondary wound together, turn for turn, with very close spacing. There is close coupling and proportionately large power transfer.

bifilar winding.—A form of *non-inductive winding* in which the current is carried through two conductors, flowing first one direction and then returning in the opposite direction around the winding.

bilateral antenna.—An antenna which radiates or receives more effectively in two diametrically opposite directions than in any other directions.

Billi condenser.—A *variable condenser* having as plates two metal tubes, one sliding within the other, with the dielectric between them.

binaural effect.—The effect of sounds on both ears, by which it is possible to determine the direction from which the sounds are coming.

binaural sound.—A sound system in which the outputs of two microphones are separately transmitted and amplified, then reproduced from two speakers placed to affect both ears of listeners. A term sometimes applied to reproduction of a single audio signal by means of two speakers producing simultaneous sets of air waves from two locations.

binder.—Liquid insulating material which, upon drying, acts as a protective coating or as a fastening for the windings of coils.

bipolar.—Having two magnetic poles.

birdies.—A name for *pip markers.*

black faced tube.—A television picture tube in which the phosphor material is darkened to reduce reflections from external sources of light.

blacker than black.—In a composite television signal, any amplitude greater than that of the black level, or any amplitude which should result in cutoff of the picture tube electron beam to leave the viewing screen dark.

black level.—In a composite television signal, an amplitude between 72½ and 77½ per cent of peak carrier amplitude. Lesser amplitude should permit a picture tube electron beam and illumination of the viewing screen, while greater amplitudes should cause beam cutoff and a dark screen. Picture signal amplitudes are less than the black level, while sync pulse amplitudes are greater than the black level.

black light.—A name sometimes applied to *ultra-violet rays*.

blanketing.—The effect of a powerful radio signal which makes reception of weaker signals difficult or impossible.

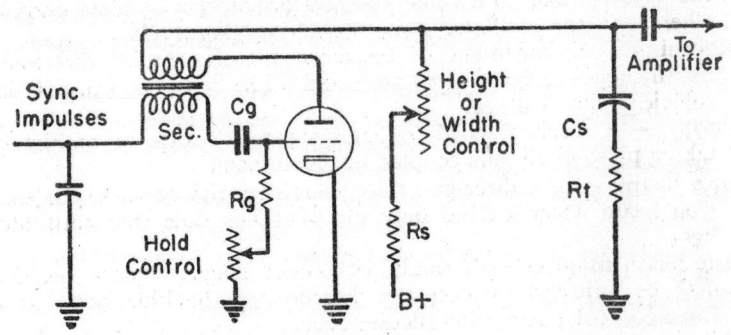

Blocking Oscillator

blanking bars.—Broad dark bars or bands appearing in television pictures. A bar from top to bottom is caused by horizontal blanking intervals, with the right-hand side of pictures at the left and the left-hand side at the right of the bar. A bar from side to side results from vertical blanking, with the top of a picture below the bar and the bottom of the picture above the bar.

blanking interval.—The time between successive horizontal lines or between successive fields and frames of television pictures. Horizontal blanking occurs between traced luminous lines, and vertical blanking between fields and frames.

blasting.—The result of tube *overloading* on the sounds emitted by a loud speaker.

bleeder current.—The steady current which flows through all sections of a power unit *voltage divider* and assists in improving the voltage regulation of the system.

bleeder resistance.—That portion of the resistance in a power unit *voltage divider* which determines the value of *bleeder current*.

blind spot.—A space within which signals from a transmitter are received with little strength or not at all.

BLOCKED IMPEDANCE

blocked impedance.—The *terminal impedance* of an acoustic device when the mechanical portion has an infinite impedance.

blocking.—Reduction of a tube's plate current to zero because of an excessive negative charge on the control grid, the circuit conditions being such that the negative charge is neutralized very slowly.

blocking condenser.—A condenser used to keep direct currents out of a circuit. A *stopping condenser*.

blocking oscillator.—As oscillator circuit with which the peak of every positive alteration of oscillating voltage charges a capacitor which is in series with the control grid of the tube. Capacitor voltage negatively biases the grid to a value greater than that for plate current cutoff. Current remains cut off until the capacitor discharges, through a grid resistor, sufficiently to lessen the negative bias to a value allowing resumption of plate current, whereupon the oscillator goes through another cycle.

blooming.—Enlargement of television pictures in all directions, usually when brightness is increased. The common cause is insufficient ultor voltage.

blooper.—A receiver which radiates signals because of *oscillation* taking place in circuits coupled to the antenna.

blue beam.—In a three-gun color television picture tube, the electron beam which excites only the phosphor dots that emit blue light.

blue beam magnet.—A small permanent magnet whose position may be adjusted to alter the direction of the blue beam in a three-gun color television picture tube.

blue glow.—The appearance of the space within a vacuum tube in which gas is undergoing ionization, due to the use of excessive voltages or to an imperfect vacuum and residual gas in the tube.

blue gun.—In a three-gun color television picture tube, the electron gun whose beam excites only the phosphor dots emitting blue light.

B-minus.—A negative *B-voltage*.

bobbin.—A winding of small size.

body capacity.—The *electrostatic capacity* which exists between parts of a person's body and radio circuit parts.

body leakage.—*Volume leakage*.

bolometer.—An instrument for measurement of heat by the change of resistance resulting from change of temperature in a conductor.

bolometer bridge.—An instrument for measuring small high frequency currents according to the change in resistance produced in a conductor which is heated while carrying these currents. The resistance element is placed in one arm of a *Wheatstone bridge*. *See illustration, page following*.

bombardment.—*Electron bombardment*.

bonding.—Connecting metallic parts together so that they form a continuous electrical conductor.

BOOSTED B-VOLTAGE

boosted B-voltage.—In television receivers, a direct voltage resulting from combination with B-voltage from the power supply of the average value of voltage pulses coming through the damper tube from the horizontal deflecting coil circuit. The pulses are partially or wholly smoothed by filtering. The boosted voltage may be several hundred volts higher than positive B-voltage.

booster.—A carrier-frequency amplifier, usually a self-contained unit, connected between antenna or transmission line and a television or radio receiver.

bootstrap voltage.—The *boosted B-voltage*.

bound charge.—An *electrostatic charge* so affected by another nearby charge that it does not escape to earth nor have an effect on an electroscope. A charge of one polarity remaining on an insulated conductor. *See illustration*.

bound electron.—An electron associated with the *positive nucleus* of an atom, not detached during electrical actions.

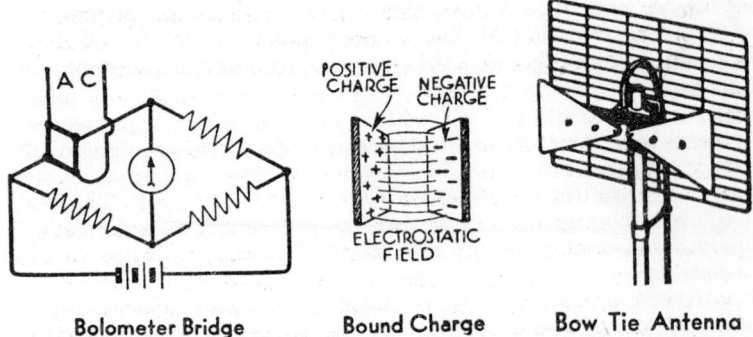

Bolometer Bridge   Bound Charge   Bow Tie Antenna

bow-tie antenna.—An antenna consisting essentially of two triangular metallic plates, solid or perforated, supported in a single vertical plane or sometimes with slight inclination forward of the outer edges. The sharper points of the plates are close together, forming a gap for connection of a transmission line. Ordinarily used for ultra-frequency reception. *See illustration*.

B-plus.—A positive *B-voltage*.

breakdown voltage.—1—The applied voltage at which an insulator or dielectric is punctured. 2—In a *grid-glow tube,* the value of grid-cathode voltage at which a glow discharge commences.

break-up.—In color television pictures, incorrect hues which appear on and near objects in rapid motion.

breathing.—Alternate expansion and contraction of television pictures because of voltage fluctuations or interference.

bridge circuit.—Four electrical elements of similar or different kinds connected together in a continuous series circuit. Each element is called an arm. An energy source is connected to two opposite or alternate junctions, and an indicator or some kind of load to the two intervening junctions. When the ratio of impedances or resistances in one pair of arms across the source equals that ratio in the other pair no current flows in the indicator or load, and the bridge is said to be balanced. See *Wheatstone bridge*.

bridge rectifier.—A circuit allowing the use of four *half-wave rectifier* elements operating as a *full-wave rectifier*. There are four arms, each containing one rectifying element, and connected in the form of a rectangle. Alternating current applied across opposite corners of the rectangle results in rectified current between the remaining two corners. *See illustration.*

brightness.—The appearance of an object which is emitting or reflecting light. See *intrinsic brightness*. Measured in *lamberts*, or in apparent *candlepower* per unit of area, or in *lumens* per unit of area.

brightness channel.—The *luminance channel* in a color television receiver.

brightness control.—A television receiver control for varying the average illumination of the picture. Also a circuit which automatically maintains a constant average illumination of the picture.

brilliance.—A quality of reproduced sound in which high frequencies and overtones have their correct emphasis in relation to the lower frequencies.

British thermal unit.—A unit for measurement of heat. The heat required to raise the temperature of one pound of water, at its point of maximum density, one degree Fahrenheit. Equal to 252 *calories*.

broad band amplifier.—An amplifier providing approximately uniform gain throughout a range of frequencies in which the ratio of highest to lowest is large. A broad band video amplifier has a ratio of about 40,000 to 1.

broad band antenna.—An antenna capable of delivering signals of approximately uniform strength throughout a range of frequencies having a high to low ratio of 2-to-1 or more.

broad tuning.—The condition existing when a tuned circuit or circuits respond not only to frequencies within one band or channel but also to a considerable range of frequencies on each side. Having little *selectivity*.

broadening resistor.—A fixed resistor in parallel with a tuned circuit to widen the resonance characteristic.

brush discharge.—A visible glow due to *ionization* of the air at edges and corners of a conductor operating at high voltage, such as the plate of a condenser. The *corona*.

brush loss.—The power dissipated in forming a *brush discharge*.

brute force filter.—A *low pass filter* depending on large values of inductance and capacity rather than on any resonance effects to oppose passage of alternating current through it. *See illustration.*

B.T.U.—Abbreviation for *British thermal unit.*

bucking coil.—An induction coil or winding so connected that its field opposes the field of another winding, their polarities being opposite.

buffer amplifier.—An amplifier tube and associated circuits for preventing undesirable reaction between preceding and following circuits. A buffer may isolate an oscillator from a load which could affect oscillation frequency, or may prevent undesired modulation, or may serve other purposes.

buffer condenser.—A condenser connected between the anode and cathode of a *cold cathode rectifier* for the purpose of lessening voltage surges and improving the wave form of the rectified current.

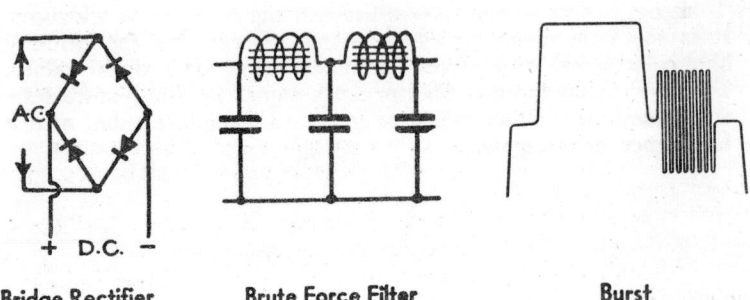

Bridge Rectifier   Brute Force Filter   Burst

burnout.—An open circuit caused by such overheating of a conductor as to cause its melting and breaking apart.

bulk eraser.—A device providing strong alternating magnetic fields for demagnetization or erasing a tape or wire recording from an entire reel without unwinding.

burst.—The portion of a color television composite signal which allows synchronizing the frequency and phase of the receiver color oscillator with those of the transmitter master oscillator. The burst consists of not less than eight sine-wave cycles at a frequency of 3.579545 megacycles per second appearing on the back porch following horizontal sync pulses. *See illustration.*

burst amplifier.—A color television receiver amplifier that selects and strengthens bursts before they are applied to the color oscillator system.

burst gate.—A color television receiver tube allowed to conduct only during periods within which bursts are received. The tube is gated to pass only the bursts.

burst oscillator.—A name for color oscillator.

burst pedestal.—Average amplitude of burst voltage with respect to the level of the horizontal blanking pedestal.

burst signal.—The *burst* in color television.

bus bar.—A rod of copper, bronze or brass used to carry large currents or to form common connection between several circuits.

bus wire.—An uninsulated tinned copper wire, square or round in section and usually of number 14 gage.

bushing.—An insulating tube or washer which carries a conductor through an opening.

butterfly antenna.—A *bow-tie antenna*.

B-voltage.—The direct voltage, from a low-voltage power supply, used for plates, screens, grid biasing, and other purposes requiring direct voltage or current in receivers, amplifiers, instruments, and other devices.

bypass condenser.—A condenser providing a low impedance alternating current path around a resistor or other part in a circuit.

B-Y signal.—One of the color-difference signals in color television. It is the blue-minus-luminance signal representing the primary blue color signal without the luminance signal or Y-signal values. The B-Y signal forms a blue primary signal for the picture tube when combined, either inside or outside the picture tube, with a luminance or Y-signal.

C

C.—The symbol for *capacitance* or *capacity* in farads.

c.—Abbreviation for the prefix *centi-*.

cabinet resonance.—The condition in which the confined air or the walls of an enclosure vibrate naturally at a certain frequency, thus producing an apparent strengthening of sounds at the resonance frequency.

cabinet speaker.—A loud speaker enclosed within a cabinet separate from the housing of the receiver or amplifier.

cable.—A number of separate conductors insulated from one another and carried within one outside covering.

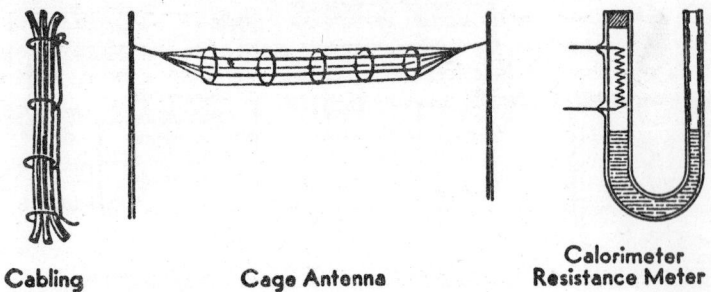

Cabling   Cage Antenna   Calorimeter Resistance Meter

cabling.—Binding together of separately insulated conductors into one cable. *See illustration.*

cage antenna.—An aerial formed by a number of parallel conductors evenly spaced around the points of a circle and connected together at their ends. *See illustration.*

calibrate.—1—To ascertain by measurement or comparison with a standard, variations between true values and actual readings of a measuring instrument. 2—To determine and record the settings of a control with reference to frequency or some other characteristic.

calido.—A resistance metal consisting of nickel, chromium and iron.

call letters.—A series of letters, or of letters and numbers, which identifies a transmitting station.

calorie.—The heat energy required to raise the temperature of one gram of water one degree centigrade. Approximately equal to 4.18 *joules,* or to 0 00397 *British thermal units.*

calorimeter instrument.—A resistance or current measuring device in which expansion of air around a conductor heated by a measured current causes a column of liquid to move upward in a U-shaped tube connected to the chamber containing the conductor and the heated gas. *See illustration, page preceding.*

cam switch.—An electric switch in which the contacts are closed by pressure exerted by a cam-shaped member. *See illustration.*

cambric tubing.—A hollow insulating tube made from oil impregnated cotton and linen cloth. *Spaghetti tubing.*

can.—1—A *shield*. 2—Slang expression for a *headphone*.

canal rays.—Beams of positively charged particles which pass away from an anode and through openings in a cathode of a vacuum tube.

candle.—An *international candle*.

candlepower.—The luminous intensity measured in *international candles* or *standard candles*.

caoutchouc.—A name for rubber.

capacitance.—1—*Electrostatic capacity*. 2—Sometimes used as meaning *capacitive reactance*.

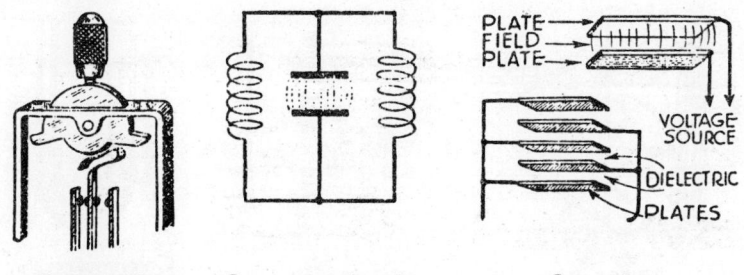

Cam Switch   Capacitive Coupling   Capacitor

capacitive circuit.—A circuit in which *electrostatic capacity* is the principal electrical factor present, the inductance being very low or negligible in comparison.

capacitive coupling.—A form of coupling in which a single *electrostatic field* is common to two circuits. Coupling by means of a capacity contained in each of the two coupled circuits. *See illustration.*

capacitive feedback.—A feedback through a *capacitive coupling* between output and input circuits.

capacitive load.—A load circuit in which the *capacitive reactance* exceeds the *inductive reactance*. A load in which the current leads the voltage. Compare *inductive load*.

capacitive reactance.—The part of the *reactance* which is due to capacity in an alternating current circuit. Measured in ohms. Equal to the number of ohms resistance which would have the same effect in opposing current flow as actually results from a condenser or capacity in the circuit. The symbol is X_c.

capacitive time constant. —The time, following the instant of connection to a voltage source, for a capacitor to charge through series resistance to 63.2 per cent of maximum which would be reached after an indefinitely long period. It is a number of seconds equal to the product of capacitance in microfarads and circuit resistance in megohms. The time is independent of applied voltage. The same time is required for a capacitor to discharge through resistance to 36.8 per cent of its initial charge.

capacitor.—A device providing capacitance in compact form. Two conductors or sets of conductors, called the plates, separated by a non-conductor called the dielectric. A difference of potential applied to the plates charges the capacitor, placing the dielectric under strain and producing between the plates an electrostatic field in which energy is stored.

capacity bridge.—A *Wheatstone bridge* for measurement of capacity; sometimes arranged to balance out the resistance and inductance of a circuit to be measured. *See illustration.*

capacity coupling.—*Capacitive coupling.*

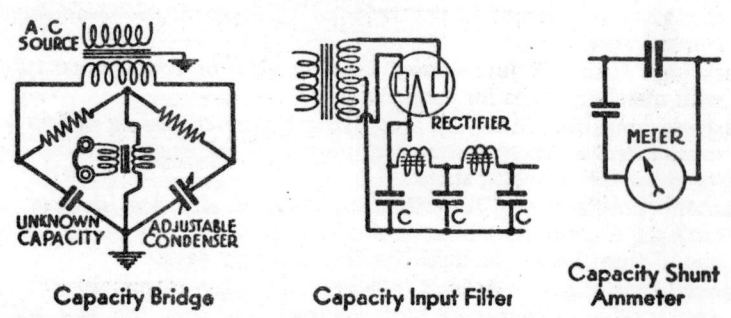

Capacity Bridge   Capacity Input Filter   Capacity Shunt Ammeter

capacity ground.—A *counterpoise.*

capacity input filter.—A *power unit* filter of low pass type in which a condenser is connected directly across or in parallel with the rectifier output. Compare *choke input filter. See illustration.*

capacity meter.—1—An instrument utilizing a *Wheatstone bridge* circuit arranged for direct measurement of *electrostatic capacities.* 2—An instrument indicating values of *electrostatic capacity* by comparison of known and unknown capacitive reactances.

capacity reactance.—*Capacitive reactance.*

capacity shunt ammeter.—A *high frequency instrument* in series with a small (high reactance) capacity, the two being in shunt or parallel with a large (low reactance) capacity placed in series with the circuit carrying the measured current.

carbon.—A conducting material occuring as graphite, lamp black, plumbago, etc. It is one of the elements. Carbon in various forms has a resistance from f.fty to several hundred times that of copper.

carbon microphone.—A microphone in which electrical resistance is varied by pressure from a sound diaphragm exerted upon a mass of loosely held carbon particles which are compressed or released by movement of the diaphragm. A *double button microphone* or a *solid back microphone*.

carbon resistor.—A fixed or adjustable resistor in which the resistance element is carbon, graphite, or a composition, and therefore is non-inductive.

carcel.—A French standard of *luminous intensity* equal approximately to 9.615 *international candles*.

carrier current.—High frequency currents modulated by a signal and carried by wire conductors. The current which exists in connection with a *carrier wave*.

carrier current telephony.—*Wired radio*.

carrier frequency.—The frequency of the original unmodulated wave of a transmitter.

carrier signal.—Modulated or unmodulated electromagnetic waves radiated from a transmitter and picked up by receivers.

carrier suppression.—A system of transmission in which the carrier wave or current is not radiated or transmitted. *Side band transmission*.

cartridge fuse.—A *fuse* carried inside a glass or composition tube with metal end caps for connection.

cascade amplification.—An amplifying system consisting of two or more *stages of amplification*, the output of one stage forming the input for the following stage.

cascade connection.—A series connection of electrical devices or circuits. A connection of several devices such that the output from one of them forms the input for the following device.

cascode amplifier.—A high-frequency amplifier circuit employing two triodes so connected that electron flow from the first plate passes to the second cathode, and from the second plate to the power supply. The input signal is applied to the grid of the first triode, connected as an ordinary grounded cathode amplifier. The second triode is connected as a grounded grid amplifier, with input to its cathode from the plate of the first triode. The combination provides high transconductance, low noise, and freedom from regenerative feedback through the tubes.

cathode.—The electrode or element at which electrons enter an electrical device from external circuits. In a filament type tube the filament is the cathode. In a battery, rectifier, or other source, the positive terminal is the cathode.

cathode bias.—Negative grid bias obtained from voltage drop across a resistor in series with the cathode of a tube, through which electron flow is to the cathode. The grid is conductively connected to the far end of the resistor, which is negative with respect to the cathode.

CATHODE FOLLOWER

cathode follower.—A tube circuit with which the plate is grounded for signal frequencies through a capacitor having small reactance at such frequencies. Input signals are applied between control grid and ground, while output is from across a resistor between cathode and ground. There is no voltage gain, rather a slight loss. The circuit allows matching a high-impedance input to a low-impedance output over a wide frequency range.

cathode heating time.—The number of seconds elapsing between application of heater voltage and a plate current equal to ninety per cent of the final value.

cathode particles.—Electrons emitted from a *cathode*.

cathode ray.—A stream of electrons emitted from a *cathode*.

cathode ray oscillograph.—An instrument using a *cathode ray tube* to make visible the wave form of a changing current or volt-

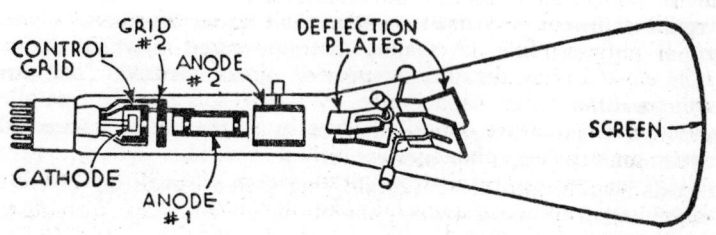

Cathode-ray Tube

age through deflection of the ray by electromagnetic or electrostatic fields produced by the current or voltage.

cathode ray tube.—A vacuum tube in which beams of electrons emitted from a cathode and given velocity by a charged anode are deflected by a magnetic or electric field and made visible by striking against a *fluorescent* screen in the end of the tube. Electric forces controlling the deflecting fields are indicated by movement of the beam. *See illustration.*

cation.—A *positive ion;* an ion which moves toward the cathode during electrolysis. See *anion.*

cat whisker.—A small wire or metallic point making contact with the mineral element in a *crystal detector.*

C-battery.—A battery which provides *negative bias* voltage for the grid of a vacuum tube.

center frequency.—In any frequency-modulation system, the frequency below and above which are approximately equal shifts or deviations during modulation.

center tap.—A connection to the electrical center of a winding, or to a point midway between the electrical ends of a resistor or other portion of a circuit. *See illustration.*

centering control.—A television receiver control for shifting the picture on the screen. The horizontal control moves the picture sideways; the vertical control moves it up or down.

centi-.—A prefix meaning one one-hundredth of the unit.

centigrade temperature scale.—One of the scales used for measurement of temperature. Zero is taken as the temperature of melting ice and the "100" point is at the temperature of boiling water.

centimeter.—1—A unit of length in the metric system of measurements; equal approximately to 0.3937 inch. 2—A C.G.S. unit of *electrostatic capacity*, approximately equal to 1.11 micro-microfarad or 1/900000 microfarad. 3—A C.G.S. unit of *self-inductance* equal to the one-thousandth part of a microhenry. Abbreviated *cm*.

centimeter-gram-second units.—Units based on the centimeter for length, the gram for mass and the second for time, these being the fundamentals in an internationally accepted system of measurement. Abbreviated *C.G.S.*

ceramic capacitor.—A fixed or adjustable capacitor whose dielectric is porcelain-like or ceramic material mixed with substances such as titanium dioxide to provide great dielectric constant and resulting large capacitance in small space. The metallic plates most often are deposited directly on the ceramic material.

ceramic pickup.—A phonograph pickup cartridge whose output voltage and current result from piezo-electric effect in a crystalline element of ceramic or porcelain-like structure.

ceresin wax.—A prepared mineral wax of high dielectric strength and resistivity. Dielectric constant 2.5.

C_f.—Symbol for cathode capacitance.

C_g.—Symbol for grid capacitance.

C_{gf}.—Symbol for grid-filament capacitance.

C_{gk}.—Symbol for grid-cathode capacitance.

C_{gp}.—Symbol for grid-plate capacitance.

C.G.S.—Abbreviation for centimeter-gram-second units.

C.G.S. line of force.—*A magnetic line of force.*

chain broadcasting.—Broadcasting from several transmitters at once of a program received over land lines from a central *key station*.

channel.—*1*—A limited range of frequencies within which a radio or television transmitter is permitted to radiate signals. *2*—In a receiver or other device handling signal frequencies, all circuits and parts intended primarily to carry one signal or kind of signal, although other signals may follow the same path to some extent.

characteristic.—A graph or curve showing the manner in which one value is caused to change as a result of changes in some other value or factor. *See illustration, page following.*

characteristic impedance.—The *iterative impedance.*

charge—1—The electricity held in a condenser or other capacity.

CHARGED BODY

Electrostatic charge. 2—Ability of a storage battery to produce an electric current.

charged body.—A body which has positive or negative *electrification*, which is carrying a *positive charge* or a *negative change*.

chassis.—The supporting metal base together with attached parts and circuit connections of a receiver or other electrical device. Parts such as speakers and picture tubes may not be part of the chassis.

chassis ground.—A metal chassis when considered as of ground or zero potential with reference to circuit elements and potentials above or below ground.

choke coil.—1—A coil having sufficient *self-inductance* to greatly diminish flow of alternating currents through it while allowing comparatively free passage for direct current. A coil which develops sufficient *reactance* to undesired frequencies to practically prevent their flow in parts of a circuit containing the coil.

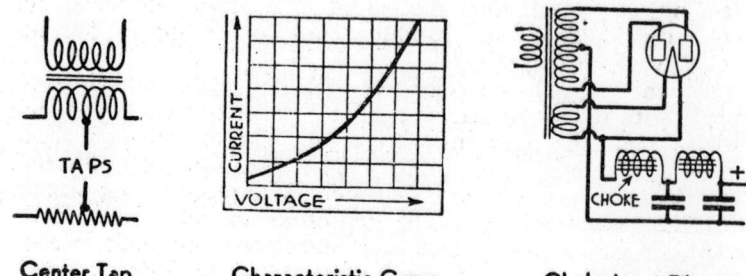

Center Tap   Characteristic Curve   Choke Input Filter

choke input filter.—A power unit type of *low pass filter* in which a choke coil is in series with the rectifier output before this output reaches the first condenser in parallel across the filter. Compare *capacity input filter*.

chopper.—A device for interrupting *continuous wave* signals at audio frequency either in the transmitter or receiver. A rotating wheel with alternate conducting and insulating segments traveling under a contact brush. A rotating variable condenser in a tuned circuit.

chroma.—In color television, a word having much the same meaning as chrominance. Refers to circuits and parts handling color or chrominance signals as distinguished from those handling brightness or luminance signals. Chroma sometimes refers only to color saturation instead of to all chrominance effects.

chroma control.—Usually the saturation control for a color television receiver.

chromatic.—Pertaining to color.

chromatic aberration.—An effect which causes refracted white light

chromaticity.—The combined effects of hue and saturation in determining appearance of a color, without considering the effects of brightness or luminance.

Chromatron.—A type of single-gun color television picture tube. The Lawrence tube.

chrominance.—Descriptive of those circuits and parts in a color television receiver which carry signals that determine color hue and saturation, with the exception of circuits controlling color synchronization. Chrominance is coloring, in the form of hue and saturation, added to gray tones of any given brightness to produce a picture with coloring instead of one in black, white, and shades of gray.

chrominance amplifier.—Amplifying and frequency-selective circuits of a color television receiver which strengthen and pass the chrominance signals.

chrominance channel.—Circuits of a color television receiver extending, in general, from the output of a video detector to the color demodulators, and handling chrominance signals but not color synchronizing signals.

chrominance signal.—In color television, the signal voltage whose phase with respect to master oscillator phase represents hue, and whose amplitude represents saturation. The chrominance signal is formed originally from primary color signals or color-difference signals, and in the receiver is demodulated to recover these signals.

chrominance subcarrier.—The *subcarrier* in color television.

circuit.—A path through which may pass electric, magnetic or electrostatic effects. The entire path traversed by an electric current, including the source, the energy consuming devices and all conductors between them.

circuit breaker.—A form of switch that will open a circuit carrying a large current without harm to the switch contacts.

circular loom.—A circular tubing composed of fabric and insulating composition which is used to insulate and protect wires.

circular mil.—A unit of cross sectional area of wires and other conductors. The area of a circle having a diameter of one mil or the one-thousandth of an inch; equal to 0.0000007854 square inch.

circulating currents.—In a parallel resonant circuit, electron flows that carry energy back and forth between inductance and capacitance, and which remain within the resonant circuit.

Clark cell.—A *standard cell* using one electrode of mercury, another of zinc amalgam and an electrolyte containing zinc sulphate and mercurous sulphate. Its voltage is 1.434.

class A operation.—Operation of an amplifier or modulator tube with *grid bias* such that plate current variations occur only on the straight portion of the *grid-plate characteristic*, without plate

CLASS B OPERATION

current cut-off and without positive grid potential on signal peaks. The distortion limit is within five per cent.

class B operation.—Operation of an amplifier or modulator tube with grid bias such that plate current is near zero with no signal and such that the grid becomes positive on signal peaks. Distortion is reduced by filtering or by using *push-pull* or balanced circuits.

class C operation.—Operation of an amplifier or oscillator tube with grid bias such that the plate current is zero with no signal and such that *saturation* plate current occurs with positive grid potential on signal peaks. *Harmonic distortion* is reduced in connected circuits.

click method.—A method of determining *resonance frequency*. A click is heard in headphones connected to an *oscillatory circuit* when the resonance frequency of this circuit is made the same as that of another circuit coupled to the first one. The click results from starting or stopping of oscillations in the first circuit

clipper.—A diode, triode, or pentode operated with such bias and other element voltages as to remove portions of a signal which exceed certain amplitude on positive or negative alterations, or both.

close coupling.—A coupling through which a comparatively large amount of energy is transferred. Usually a coupling of which the *coupling coefficient* is 0.5 or greater. A coupling in which the *mutual inductance* of two circuits is large in comparison with their *self-inductances*.

closed antenna.—A *loop antenna*.

closed circuit.—A circuit which is complete and through which may pass electric currents or magnetic lines of force.

closed circuit jack.—A jack through which a circuit is normally closed. *See illustration*.

closed circuit voltage.—The voltage across the terminals of a source when current is flowing from it.

closed core.—An iron or steel core extending all the way around a magnetic path with the possible exception of one or more small gaps used to control the amount of flux. *See illustration*.

closed field coil.—An inductance coil having a small external field.

cm.—Abbreviation for *centimeter*.

coarse tuning.—*Broad tuning*.

coated filament.—An *oxide coated filament*.

coaxial speaker.—One large and one small speaker supported as a unit, with their axes in line. Operation is from a single input signal, with reproduction of higher audio frequencies by the smaller speaker and of lower frequencies by the larger one.

coaxial transmission line.—Cable with a central conductor surrounded by insulation, with another cylindrical conductor around the insulation. Insulation may be solid or plastic, or chiefly air except for spaced supports carrying the central conductor. The outer conductor may be solid metal tubing or flexible braid. Around the outside of the cylindrical conductor may or may not be a protective covering.

code.—A system of signals used in telegraphy. The continental code is used in radio telegraphy. It consists of combinations of dots and dashes representing letters, numerals, punctuation and complete phrases.

code rules.—The rules of the *National Electric Code*.

coefficient.—A number indicating the rate of change or the amount of change caused in some quantity or value by variation of conditions or by variation of another quantity.

coercive force.—The *magnetizing force* which must be applied in a direction opposite to the original magnetizing force in order to completely remove *residual magnetism*.

coffer.—A deeply recessed panel; useful in acoustical treatment of auditoriums.

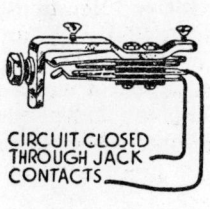
Closed Circuit Jack

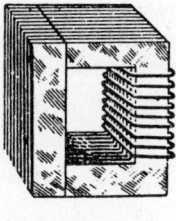
Closed Core

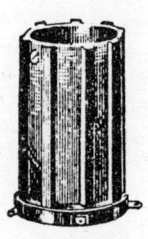
Coil Form

coil.—An *inductance coil*.

coil antenna.—A *loop antenna*.

coil form.—The insulating support upon which an *inductance coil* is wound. *See illustration.*

cold cathode grid-glow tube.—A *grid-glow tube* capable of handling only small currents in the controlled circuit, there being no primary source of electrons in the tube.

cold cathode rectifier.—A *gaseous conduction rectifier* in which ionization and conduction of current take place through a very small space between the electrodes, under the influence of applied alternating voltages and without the assistance of *primary emission* from a heated cathode.

collecting lens.—A lens used near a source of light for the purpose of collecting a quantity of light rays.

collector.—A transistor element that carries electrons from external circuits to an *n*-type base, or from a *p*-type base to external circuits. To the collector is applied a negative or positive bias, as required to aid electron flow. In a point-contact transistor the collector is one of two small, finely pointed wires, and in a junction transistor is a crystal of a type opposite to the type in the base.

color balance.—In color television, such relative intensities of the three primary colors as combine to give the visual sensation of white light.

color carrier.—The *subcarrier* in color television.

color coding.—*1*—Bands, stripes, or dots of different colors which represent certain numerical values as well as tolerances and other characteristics of fixed resistors and capacitors on which the colors are applied. *2*—Colors and combinations of colors in wire insulation to indicate particular circuits of which the wires are a part, or to which of various windings and other internal parts of electrical devices are connected the coded wires.

color-difference signal.—A color television signal equivalent to that for a primary color minus the luminance or Y-signal voltage. Color-difference signals are B-Y for blue-minus-luminance, G-Y for green-minus-luminance, and R-Y for red minus luminance. Combining a positive luminance or Y-signal with one of these produces the corresponding primary color signal.

color gate.—In a color television receiver employing a single-gun picture tube, an amplifier tube made conductive only during the period in which the electron beam is switched to a phosphor of one primary color. There is a blue gate, a green gate, and a red gate.

color grid.—In a single-gun color television picture tube, immediately back of the phosphor plate, an assembly of thin wires parallel to one another and to the phosphor strips. While adjacent wires are made of opposite polarities they produce between them electric fields that focus the electron beam away from green phosphor strips and onto either blue or red strips.

color killer.—A method for preventing appearance of color on a color television picture tube while receiving black-and-white or monochrome transmissions. A killer tube and associated circuits bias one or more chrominance amplifiers to cutoff except during reception of signals containing bursts, which are transmitted only during color broadcasts.

color oscillator.—In a color television receiver, an oscillator operating continually at the color subcarrier frequency of 3.579545 megacycles per second, synchronized with the transmitter master oscillator by bursts, and furnishing voltages that combine with chrominance signals in the color demodulators.

color phase.—Phase of chrominance signal voltage with respect to phase of the master oscillator at the transmitter or the color oscillator in the receiver. This color phase determines the hues which appear in pictures.

color sync.—The *burst* signal.

Colpitts oscillator.—A vacuum tube oscillator whose tunable resonant circuit consists of a single inductor between control grid and plate, and in parallel with the inductor two capacitances in series with each other. From between the capacitances is a connection to the tube cathode. One or both capacitances may be fixed or adjustable capacitors, or tube interelectrode capacitances, or stray and distributed capacitances in associated circuit parts. Tuning may be with adjustable capacitance or adjustable inductance. Many modifications of the basic circuit are employed. *See illustration*.

common connection.—A single connection to two or more circuits.

common logarithm.—See *logarithm*.

comparison method.—Measurement of an unknown value by the adjustment of a similar characteristic of known value so that the two bear a definite ratio to each other. Resistance measurement with a *Wheatstone bridge* is an example.

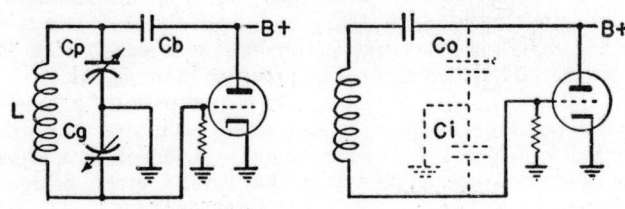

Colpitts Oscillator

compatibility. —Descriptive of a color television system that allows ordinary unaltered black-and-white or monochrome receivers to produce black-and-white pictures from color television transmission, and allows color receivers to reproduce in black-and-white pictures that are so transmitted.

compensated volume control.—A volume control circuit which attenuates middle and high audio frequencies more than low frequencies when the volume level is reduced, thus compensating for lessened sensitivity for low frequencies at low volume which is a natural characteristic of human ears.

complementary color.—Either of two colors which, when combined, cause the visual sensation of white, although the only wavelengths present are those of the two colors. Color television primaries and their complementary colors are blue and yellow, green and magenta, red and cyan.

complete modulation.—*Modulation* in which the amplitude of the carrier is caused to drop to zero. One hundred per cent modulation. *See illustration*.

complex coupling.—A combination of *inductive* and *capacitive couplings*.

complex wave.—A *periodic wave* made up of a combination of several frequencies or several sine waves superimposed on one another. *See illustration*.

compliance.—A measure of the flexibility of a mechanical part when acted upon by a force. The reciprocal of stiffness. Expressed in centimeters of movement per dyne of force.

component.—One of the several parts in a total value; a part which may be subject to independent variation in value. Any one of several different forces which, acting together, are the equivalent in effect of the single force of which they are said to be components.

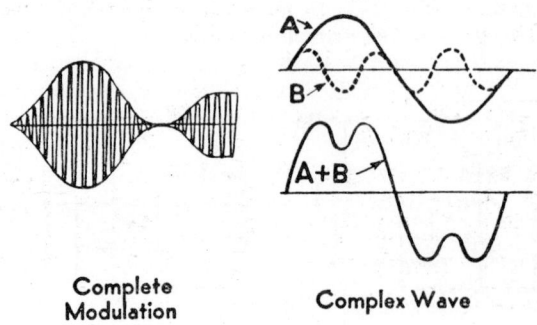

Complete Modulation

Complex Wave

composite color signal.—Amplitude modulation of the carrier for color television, including chrominance, luminance, and burst signals, together with all synchronizing and blanking signals. Accompanying sound is transmitted by frequency modulation.

composite television signal.—Amplitude modulation of the carrier for black-and-white or monochrome television, including video frequencies for pictures, and all synchronizing and blanking signals. Accompanying sound is transmitted by frequency modulation in the same channel but as a separate signal.

compression condenser.—A condenser in which the electrostatic capacity is varied by pressing the plates closer together or allowing them to spring farther apart.

compression lightning arrester.—A *lightning arrester* consisting of a resistance in series with air gaps enclosed by a tube. The arc expands and compresses the enclosed air.

compressional vibration.—*Longitudinal vibration*.

concatenation.—*Cascade connection*.

concave lens.—A lens which is thinner through the center than around the edges.

CONDENSANCE

condensance.—*Capacitive reactance.*

condenser.—A device which provides *electrostatic capacity* in compact form. A condenser consists of two conductors, or sets of conductors, called the plates, separated by a non-conducting medium called the *dielectric*. A difference of potential applied to the plates charges the condenser, placing the dielectric under strain, in which condition it stores electric energy which is reconverted into electromotive force upon discharge of the condenser.

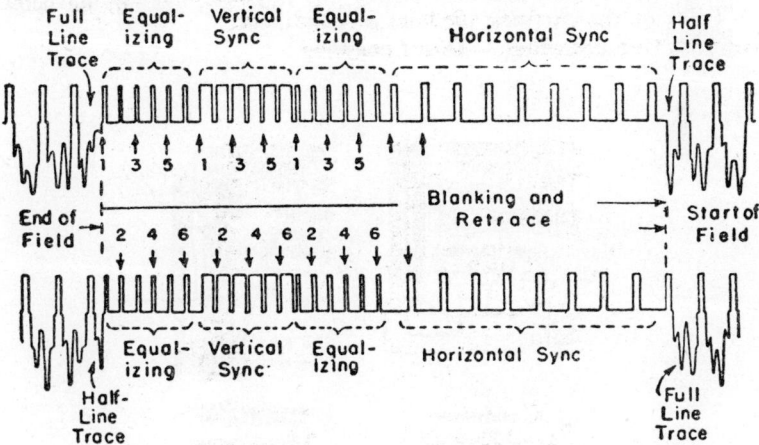

Composite Television Signal

condenser leakage.—A slow penetration of current through a condenser's *dielectric*, allowing gradual discharge or allowing a steady flow of current through the condenser, the dielectric acting as an imperfect insulator.

condenser loud speaker.—A loud speaker in which sound is radiated from a moving member forming one plate of a *condenser* in an electric circuit carrying signal voltages. The movement results from *electric attraction* which varies with the signal voltages. *See illustration.*

condenser microphone.—A *microphone* in which sound waves cause relative motion between the plates of a condenser, thereby varying the electrostatic capacity and *impedance* of a connected circuit in which are set up current variations corresponding to the sound wave. *See illustration.*

condenser pickup.—A *phonograph pickup* in which the electric output is generated by mechanically controlled changes in *electrostatic capacity*.

CONDUCTANCE

conductance.—1—The ability to carry or conduct electricity. The reciprocal of the *resistance* in a direct current circuit and the active component of the *admittance* in an alternating current circuit. Measured in *mhos*. The symbol is G or g. 2—*Photocell conductance*.

conduction.—Transmission of electricity, sound or heat through a medium without physical motion of the medium itself, as in the flow of electric current through a wire.

conduction current.—Current flowing in a conductor because of differences of potential between parts of the conductor. Current flow which is not accompanied by any physical movement between parts of the carrying medium or conductor.

conductive coupling.—*Direct coupling*.

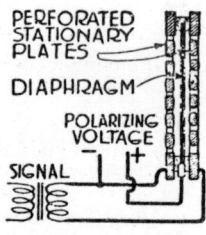
Condenser Loud Speaker

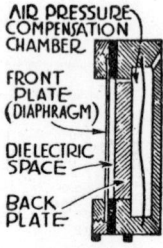
Condenser Microphone

conductivity.—The *conductance* of a material expressed in numerical units. The number of mhos conductance between opposite faces of a centimeter cube. The reciprocal of *resistivity*. Also called specific conductance. The symbol is the Greek letter gamma (γ).

conductor.—Any material which allows electric current to flow through it continuously when voltage is applied. Metals are the most common conductors. Specifically; one or more wires, not insulated from one another, and capable of carrying a single electric current.

conduit.—*Rigid iron conduit* or tubing within which are carried wires to be protected and supported. Also *flexible metal conduit* or tubing.

conical antenna.—A dipole antenna whose oppositely extending conductors form an angle of less than 180 degrees, with its center or bisecting line on the direction from which desired signal waves approach the antenna. There is increased pickup of such waves, and reduced pickup of waves approaching from the opposite direction. *See illustration.*

conical horn.—A horn in which the equivalent sectional radius increases at a constant rate.

conjugate foci.—Two points on opposite sides of a lens, at either one of which may be placed a point source of light with its image appearing at the other point. The source and the image are interchangeable in position between the two conjugate foci.

connected load.—The sum of all the *continuous ratings* or loads connected to a system.

conservation of energy.—The principle that energy cannot be created nor destroyed, but only can be changed from one form into another. Some of the total energy may change into a form no longer useful or available.

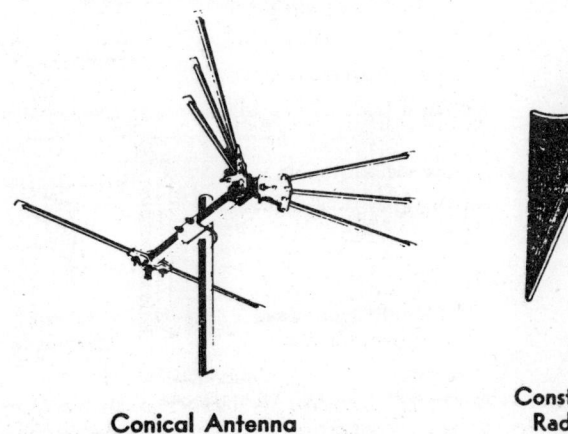

Conical Antenna   Constrained Radiator

console.—An ornamental cabinet for a receiver, the cabinet standing from the floor on legs.

consonance.—*Resonance*, either electrical or acoustical; especially as occurring in bodies or circuits which are coupled but not directly connected.

constant.—A quantity which expresses a fixed value, condition or property of a material.

constant amplitude recording.—A method of phonograph disc-recording with which movement of the cutting stylus and reproducing needle sidewise with respect to the groove is limited to the same maximum value for all frequencies, instead of increasing as frequency decreases.

constant luminance.—One of the principles of color television transmission and reception, namely, that luminance or brightness is controlled solely by the luminance signal, while color hue and saturation are controlled solely by the chrominance signal.

constant velocity recording.—A method of phonograph disc recording with which sidewise movement of cutting stylus and reproducing needle along the groove is inversely proportional to frequency, being large for low frequencies and relatively small for high frequencies.

constrained radiator.—The sound radiating portion of a loud speaker in which the original mechanical motion acts first upon a part of the air confined in a horn or similar enclosure. *See illustration.*

contact potential.—A difference of potential between a heated cathode in a vacuum tube and other elements on which electrons accumulate when no external element voltages are applied. Other elements become negative with respect to the cathode. The effect is most pronounced on the element closest to the cathode in multi-element tubes.

Contact Rectifier

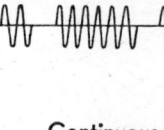

Continuous Wave Transmission

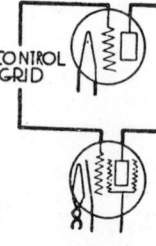

Control Electrode

contact rectifier.—A *crystal detector*, a *copper oxide rectifier*, a *sulphide rectifier* or any contact between two materials which allows flow of current more easily one way than the other between the materials. *See illustration.*

contact resistance.—The resistance across two conductors at their point of contact.

contactor.—A device which opens and closes a circuit during regular operation of a device.

continental code.—A system of telegraphic signals used generally in European countries and for radio telegraphy in America.

continuity test.—A test for determining whether a circuit is complete, or is open and incapable of carrying the kind of current it should handle.

continuous current.—A steady direct current, one which does not vary in strength and which contains no appreciable alternating component.

continuous loading.—The addition of inductance uniformly distributed along the length of a transmission line.

continuous oscillations.—*Continuous waves.*

continuous rating.—An output which may be maintained continuously without exceeding specified limits of temperature, etc.

CONTINUOUS WAVES

continuous waves.—Radio waves which maintain a constant amplitude and constant frequency. Abbreviated *C.W.* or *cw*. Undamped waves.

continuous wave transmission.—Radio telegraph transmission by *continuous waves* which are broken up into the dots and dashes of the code. *See illustration.*

contrast.—In television pictures, the range of gray tones between darkest and lightest. Full range contrast extends from black to white with no tones which should be dark gray becoming black and none which should be light gray becoming white. With insufficient contrast there are no whites or deep blacks, only grays. With excessive contrast shades which should be light gray become white and those which should be dark gray become black.

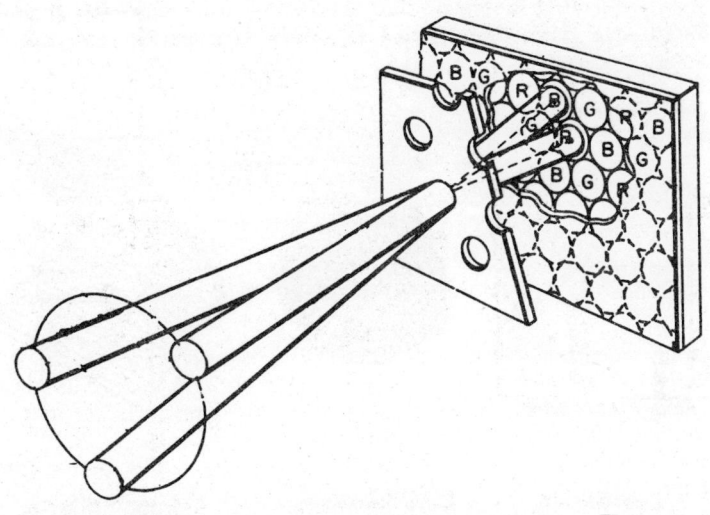

Convergence

contrast control.—A television receiver control for varying the range of illumination or the brightness difference between parts of the picture.

control grid or electrode.—A tube element whose potential with respect to the cathode regulates the rate of electron flow from cathode to plate or anode, or may prevent such flow. Making a control grid more negative decreases or prevents electron flow, while making the grid less negative or positive increases the rate of flow.

convergence.—In a three-gun color television picture tube, the coming together of electron beams from the three guns at any given

CONVERSION TRANSCONDUCTANCE

single opening in the aperture mask, with following separation or divergence in the space between mask and viewing screen so that each beam strikes only phosphor dots emitting one color. See *aperture mask*.

conversion transconductance.—The ratio of change of intermediate-frequency signal current in the output circuit of a mixer tube to the change of radio-frequency voltage at the mixer control grid, with all other element voltages unchanged.

converter for ultra-high frequencies.—A self-contained device employing double conversion for changing modulated ultra-high frequency carriers to very-high frequency modulated carriers for application to the antenna input of a very-high frequency television receiver. The converter provides the first conversion.

converter.—In a superheterodyne circuit, a single tube having elements allowing action as both oscillator and mixer, for changing modulated high frequencies to lower frequencies carrying the same modulation.

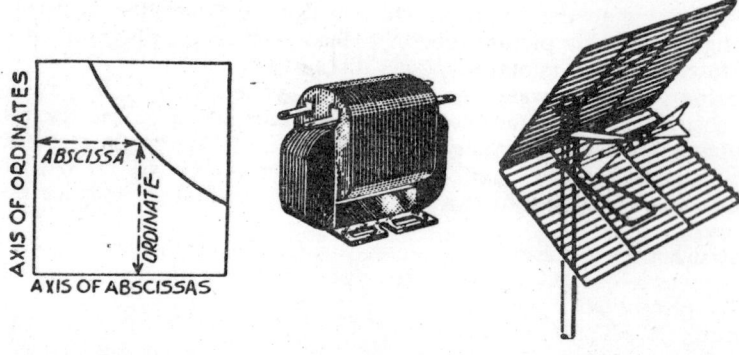

Coordinates   Core Transformer   Corner Reflector

coordinates.—*Abscissas* and *ordinates*. See illustration.
co-phasal.—Having the same *phase*.
copper.—A soft, malleable metal of reddish color, the most important and generally used of all electrical conductors. The resistivity of annealed copper at a temperature of 20° centigrade is 10.371 ohms per mil-foot and its temperature coefficient of resistance for the same temperature is 0.00393.
copper clad steel.—Wire having a steel center covered with copper.
copper loss.—The energy dissipated as heat in conductors. The loss resulting from resistance. The I^2R loss. Measured in *watts*.
copper oxide rectifier.—A *rectifier* employing a copper oxide coating on a piece of metallic copper, these two materials showing unilateral conductivity.

core.—1—The iron which forms a path for the *magnetic circuit* in a transformer, a choke coil, etc. See *transformer core*. 2—The conductor of a cable.

core loss.—The *iron loss*.

core transformer.—A transformer in which the *core* forms a single continuous ring or rectangular piece carrying the windings. Compare *shell transformer*. See *illustration*.

corner reflector antenna.—A dipole antenna with a reflector of metallic conductors arranged in two planes meeting at an approximate right angle and inclined equally upward and downward from the horizontal. The line on which the planes meet is parallel to a horizontal axis through the dipole and directly behind this axis. *See illustration*.

corona.—A pale purple light caused by ionization of the air near a body carrying high voltage or highly charged. A *brush discharge*.

corona loss.—The power dissipated in the electric discharge forming a corona.

corona ring.—A conductor, usually circular, supported near the lugs of the socket for the rectifier in a high-voltage power supply, and connected to the lug from which is connection to the high-voltage anode of a picture tube or cathode-ray tube. The purpose is to confine and minimize corona discharges.

cosine yoke.—A wide angle deflecting yoke whose coils are so shaped as to maintain nearly uniform trace height as the electron beam travels horizontally all the way across the viewing screen. There is lessened astigmatism or deformation of the beam spot.

cosmic.—Pertaining to the entire universe rather than only to the earth.

cosmic rays.—Rays of exceedingly high frequency coming from origins as yet unknown in outer space. Wavelengths less than 0.01 *Angstrom units*.

coulomb.—The practical unit of electrical quantity. The quantity of electricity passing during one second in a circuit carrying one ampere. The quantity of electricity contained in a condenser of one farad capacity when there is a potential difference of one volt between the plates. The symbol for quantity of electricity in coulombs is Q.

Coulomb's law.—A law which states that the attractive or repulsive force of two electric or magnetic charges upon each other is directly proportional to their quantities and inversely proportional to the square of the distance between them.

counter-electromotive force.—A voltage developed in an *inductive circuit* by an alternating or pulsating current, the polarity of this voltage being at every instant opposite to that of the applied voltage. The counter-electromotive force results from cutting of the circuit's conductors by the moving lines of force produced by the varying current. Abbreviated *counter-e.m.f.*

counterpoise.—Metallic conductors placed either on or a few feet above or below the ground, directly under aerial wires, and used in place of or in conjunction with the earth or ground as part of a *capacity antenna* circuit. The counterpoise is insulated from the ground. See *illustration*.

couple.—Two parts between which there is electrical action.

coupled circuit.—A circuit containing a capacity, inductance or

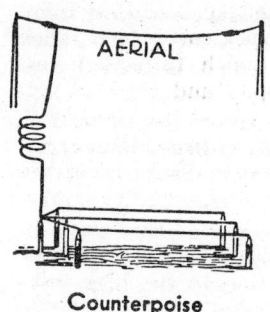
Counterpoise

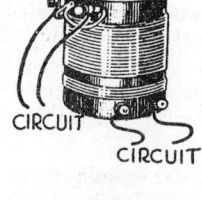
Coupler

Crevass

resistance which is also contained in another circuit so that energy may be transferred from one circuit to the other through the common element. A circuit affected by another either conductively, inductively or electrostatically. See *coupling*.

coupler.—Coils, condensers, resistances or combinations of these parts so connected that they provide coupling between two circuits. See *illustration*.

coupling.—The means by which electrical energy is transferred from one circuit to another. Coupling may be *direct, inductive, capacitive* or *resistive*.

coupling capacitor.—A capacity through which two circuits have *capacitive coupling*.

coupling coefficient or factor.—A numerical measure of the amount of coupling between two circuits. The ratio of the mutual inductive reactance, capacitive reactance or resistance of two circuits to the square root of the product of the separate similar reactances or resistances in the two circuits. The usual symbol is k.

coupling inductor.—An inductance coil or coils providing *inductive coupling* between two circuits.

coupling transformer.—A transformer used to provide *inductive coupling* between audio frequency or radio frequency circuits.

C_p.—Symbol for *plate capacitance*.
C_{pf}.—Symbol for *plate-filament capacitance*.
C_{pk}.—Symbol for *plate-cathode capacitance*.

c. p. s.—An abbreviation for *cycles per second*.
crest factor.—The ratio of the maximum value to the *root-mean-square value* in an alternating wave.
crest voltmeter.—A *peak voltmeter*.
crevass.—A pronounced dip in a *resonance characteristic*. See illustration.
critical angle.—The *angle of incidence* of a light ray which, if exceeded, results in a total *reflection* of the ray back into the medium in which it first exists and in no penetration of the ray into the medium beyond the reflecting boundary.
critical coupling.—Maximum coupling coefficient which allows only a single resonant peak, and from which further increase would cause the peak to become flat-topped and then to separate into two peaks.
critical damping.—The damping present with *critical resistance*.
critical resistance.—In an *oscillatory circuit*, a resistance just great enough to prevent *sustained oscillation* at the circuit's natural frequency. A resistance in number of ohms greater than twice the square root of the inductance in henrys divided by the capacity in farads.
Crookes' dark space.—A non-luminous region of slight depth around the surface of the cathode in a tube carrying current through an ionized gas.
Crookes' tube.—A tube containing gas under low pressure in which ionization and a luminous glow result from passage of an electric current.
cross.—A *short circuit* or accidental ground caused by two conductors coming in contact where they cross each other.
crosshatch pattern.—Regularly spaced vertical and horizontal lines intersecting to form right angles on the viewing screen of a television picture tube.
cross modulation.—In a radio frequency amplifying tube having a high negative bias and simultaneously affected by a strong signal and a weaker one, a rectification effect which causes changes of average plate current at the *modulation frequencies* of the strong signal. The weaker signal is amplified in the usual manner, but the variations in average plate current carry the effect of the strong signal along with that of the weaker one through the amplifier and detector where both signals are made audible.
crossover frequency.—With two or more speakers fed with a single audio signal, the frequency below which most of the input power goes to a larger speaker for reproduction of low tones and above which most of the power goes to a smaller speaker for reproduction of higher tones. Crossover is caused by a filter.
cross talk.—Speech or sounds in one telephone circuit carried into another circuit by *electromagnetic* and *electrostatic induction* between the conductors.
CRT.—Abbreviation for cathode-ray tube.

CRYSTAL

crystal.—A mineral body having a definite internal structure which results in the external surfaces being plane and arranged symmetrically.

crystal control.—See *piezo-electric*.

crystal detector.—A detector using a crystal diode. Formerly a detector using an exposed crystal of galena or other mineral with an adjustable catwhisker.

crystal diode.—A rectifier in which a small crystal of germanium, silicon, or certain other minerals is contacted by the tip of a fine wire conductor. There is little opposition to electron flow from the crystal, acting as a cathode, to the wire which is the anode, but relatively great opposition to opposite flow. The entire structure is enclosed, and there are no adjustments. Rectifying efficiency remains good at very-high and ultra-high frequencies, depending on the kind of crystal.

crystal oscillator.—A vacuum tube oscillator whose operating frequency is held within close limits by employing a piezo-electric crystal of quartz in the control grid circuit of the tube.

crystal pickup.—A phonograph pickup cartridge in which a piezo-electric crystal, often of Rochelle salt, changes needle movement into electric voltage current.

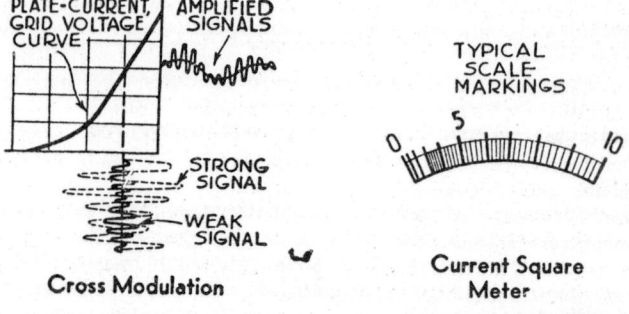

Cross Modulation   Current Square Meter

cuprous oxide cell.—A device exhibiting photovoltaic action. A cuprous oxide cathode and an inert anode are immersed in an electrolyte sealed within a glass container. See *photo-voltaic*.

Curie cut.—An *X-cut* for a quartz crystal.

current.—The *electric current*.

current amplification.—The ratio of the signal current in the output circuit of an amplifier to the signal current in its input circuit.

current density.—The amount of electric current passing through a given cross section or area of a conductor, such as so many amperes per square inch. The ratio of the current in amperes to the cross sectional area of the conductor.

current feed.—A connection from the closed circuits of a transmitter to its antenna at a point where the antenna carries a large current at low voltage.

current feedback.—A method of degeneration with which feedback voltages result from signal currents in a resistor, usually a resistor in series with the tube cathode.

current flow.—Movement of free electrons. This movement is away from any point that is negative, due to excess electrons, toward another point relatively positive, due to deficiency of electrons. Earlier it was assumed that electricity flows from positive to negative. Although there is no such flow, the erroneous assumption still is followed in some fields of electricity.

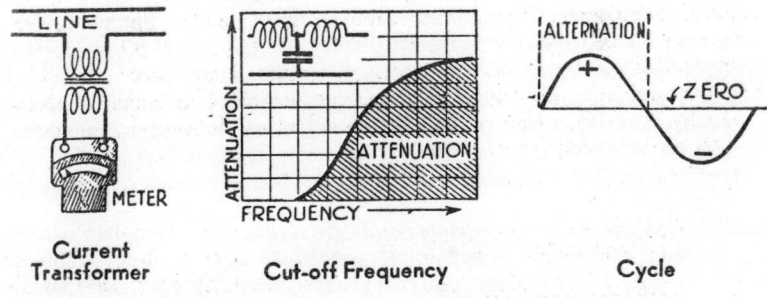

Current Transformer   Cut-off Frequency   Cycle

current limiting reactance.—A coil of which the *reactance* reduces the current through an alternating current circuit in which there exists a short circuit or other overload.

current ratio.—The ratio of the effective primary current to the effective secondary current of a transformer.

current regulator.—A coil in which the reactance or inductance is automatically changed to maintain a constant current in a circuit with varying loads.

current relay.—A relay which operates upon a change in current value.

current resonance.—The condition of *parallel resonance*.

current square meter.—A current measuring device utilizing the effects of heat, such as a *hot-wire* or a *thermocouple instrument*, in which movement of the pointer is proportional to the square of the current flowing. *See illustration.*

current transformer.—1—A transformer having its primary in series with a current carrying conductor and its secondary connected to a meter calibrated to measure the current in that conductor. 2—A transformer used to prevent mutual effects between inductances and capacities in a meter and those in a circuit carrying a current measured by the meter. *See illustration.*

cushion socket.—A vacuum tube socket which allows the tube to move rather freely within narrow limits so that vibrations at high frequency are quickly damped and do not greatly affect the tube's internal elements.

cutoff frequency.—1—The frequency at which a *filter* system or *attenuation network* changes its characteristic more or less sharply, either facilitating or opposing the passage of frequencies higher or lower than that termed the cutoff frequency. *See illustration.*
2—The frequency above or below which a tube, a line, an amplifier, a loud speaker, a microphone or any transducer ceases to function efficiently.

cutter.—That portion of a sound picture disc-recording apparatus which cuts the sound grooves in the original disc, or *wax master*.

C.W. or cw.—An abbreviation for *continuous waves*.

cyan.—A blue-green color that is the complementary of red. In color television chrominance signals, cyan is 180 degrees away from red, and lags the reference phase by 256.5 degrees.

cycle.—1—One complete positive *alternation* and one complete negative alternation of an alternating current. The current starts in one polarity, rises to maximum value, falls to zero, reverses, rises to maximum value in the opposite polarity and returns to zero value. *See illustration.* 2—A complete set of any recurrent values.

cylindrical faceplate.—A television picture tube faceplate whose exposed surface is approximately that of part of a cylinder of very large radius. The only curvature is from side to side, while a vertical line anywhere on the faceplate is straight from top to bottom.

D

D.—Symbol for *electrostatic flux density*.
d.—Abbreviation for the prefix *deci-*.
damped alternating current.—A current of which successive amplitudes progressively decrease. *See illustration.*
damped impedance.—The impedance at the terminals of the electrical or the mechanical system of an *electro-acoustic transducer* when the impedance of the other system is infinite.
damped oscillation or wave.—One of a series of waves in which the *amplitude* of successive cycles is progressively smaller.

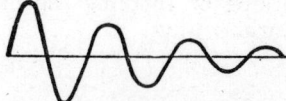
Damped Alternating Current

Damped Wave Transmission

damper tube.—A heavy-duty diode having either its plate or cathode connected to one side of the horizontal deflecting coil circuit in a magnetic deflection system. The other element of the damper connects, usually through an inductor, to positive B-voltage circuits. At the end of the first half-cycle of oscillation at natural frequency in the deflecting coil circuit the damper becomes conductive and loads the circuit to cause gradual decrease of oscillatory current instead of continuing oscillation.
damping.—1—Reduction of amplitude or strength of a radio wave or of a sound wave due to absorption and dissipation of power with time, distance traveled, and character of mediums in the path of the wave. 2—of a circuit: The progressive dropping off in value of voltage and current measured at one point in a line after the energy source has been removed. 3—of an instrument: Bringing the pointer to rest more quickly than with a free swing by the introduction of electrical or mechanical losses which absorb power.
damping coefficient, constant or factor.—1—The Naperian logarithm of the ratio between two successive similar values of an exponentially decreasing quantity. The product of the *logarithmic decrement* and the *frequency* in a circuit. Depends on the relative values of effective resistance and reactance entering into the total impedance of a circuit or electrical device. 2—The ratio between the angular movements of an instrument pointer in two successive swings, one in each direction.

daraf.—A unit of *elastance;* equal to the elastance of a centimeter cube of air.

dark current.—A small current which flows in a *photocell* because of applied voltage alone, and when there is a complete absence of light.

dark light.—*Infra-red rays.*

dark space.—*Crookes' dark space* or *Faraday's dark space.*

d'Arsonval meter.—A *moving coil instrument.*

db.—Abbreviation for *decibel.*

D.C., d.c. or d-c.—Abbreviations for *direct current.*

D.C.C.—Abbreviation for *double cotton covered wire.*

D.C. convergence.—*Static convergence.*

D.C. restoration.—Automatic variation of average control grid voltage in a television picture tube to make average brightness of reproduced pictures the same as average brightness in televised scenes. Alternating video signals originally superimposed on a direct voltage representing average brightness lose this direct component in passing through a blocking or coupling capacitor. The d-c component is reinserted in various ways.

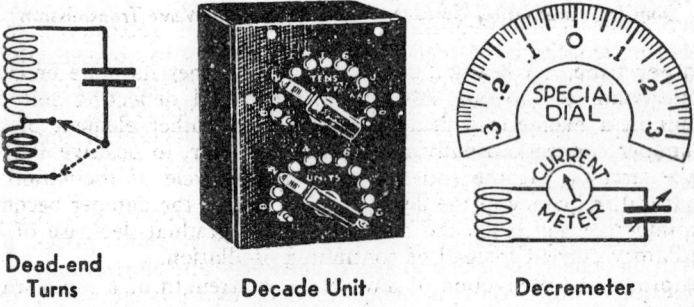

Dead-end Turns   Decade Unit   Decremeter

dead.—1—Having a *reverberation period* too short for realistic effect of sounds. 2—Not connected to a source of electromotive force; at *ground potential.*

deadbeat instrument.—A meter in which the pointer comes to rest quickly, or without back and forth movement.

dead belt.—The area included within the limits of *skip distance.*

dead end switch.—A switch which disconnects the *dead end turns* of a coil from the active turns.

dead end turns.—Turns at one end of a coil which are not included in the active circuit but which are still conductively connected to the main portion of the winding. *See illustration.*

dead spot.—A region within which signals from certain transmitters are received with difficulty or not at all.

decade unit.—An adjustable resistance, capacitance, etc., consisting of several sections, each divided into ten parts, with each tenth part of a following section equal in value to the whole preceding section. *See illustration.*

decadent wave.—A *damped oscillation.*

decay.—Gradual reduction of *amplitude* in successive cycles of a periodic wave of current, voltage, etc.

deci-.—A prefix meaning one-tenth of the unit. Abbreviated *d.*

decibel.—A unit for measurement of *gain* or of *attenuation* in power, voltage or current between different circuits or different parts of a circuit. The unit for power measurements is ten times the common logarithm of the ratio between the two powers compared. For voltage and current measurements it is twenty times the common logarithm of the voltage or current ratio when the terminal impedances are equal. Abbreviated *db.*

decoupling.—Prevention or reduction of *feedback* effects.

decrement.—*Logarithmic decrement.*

decrement method.—The *reactance variation method.*

decremeter.—An instrument for measuring *logarithmic decrement.* Measurement is made according to the change required in the capacity of an *oscillatory circuit* to produce a certain change in current. *See illustration.*

de-emphasis.—In frequency-modulation sound reception, reduction of amplitude at higher audio frequencies to compensate for pre-emphasis of these frequencies at the transmitter. A de-emphasis filter is at the audio output of the f-m demodulator.

definition.—In television pictures, good reproduction of small details and sharp outlines, due to the higher video frequencies acting in the picture tube grid-cathode circuits.

deflecting plates.—In a cathode-ray tube employing electrostatic deflection, the pairs of oppositely charged metallic plates that turn or deflect the electron beam vertically and horizontally.

deflecting yoke.—With a magnetic deflection system for a television picture tube or cathode-ray tube, the structure around the tube neck near the flare within which are coils that produce electromagnetic fields for deflecting electron beams vertically with one pair of coils, and horizontally with a second pair.

deflection angle.—The total angle through which an electron beam in a television picture tube or cathode-ray tube is turned vertically or horizontally during production of pictures or traces. This angle is twice that by which the beam is turned either way from a straight line passing through the axis of the electron gun and the center of the viewing screen.

degeneration.—Feedback of signal-frequency energy from the output of an amplifier tube or system to the input, with feedback of opposite phase with respect to signals coming to the input from a preceding stage or an external source. Degeneration allows more uniform gain over a wide range of signal frequencies.

deka-.—A prefix meaning ten times the unit.

delay distortion.—In color television pictures, distortion caused by chrominance signals taking slightly more time in passing through chrominance or bandpass amplifier circuits than taken by luminance signals in their circuits. Color and brightness then are not correctly related in pictures.

delay line.—A unit containing inductance and capacitance through which signal waves travel more slowly than in simple conductive circuits. Used in color television luminance sections to slow the luminance signals by the same amount as chrominance signals are delayed in chrominance channels. Prevents delay distortion.

delayed automatic gain control.—Automatic gain control with which the small negative control voltage produced from weakest received signals is opposed by a small positive voltage. There is no reduction of gain until received signals cause a negative control voltage overcoming the positive delay voltage.

delta (δ).—Greek letter symbol for *logarithmic decrement*.

demodulation.—A process which obtains from a modulated wave the signal impressed upon the wave during modulation. The process of *detection*.

demodulator.—A *detector*.

density.—The ratio of a total quantity to the number of units of area, volume, etc., in which it is uniformly distributed; as the ratio of weight to volume.

design center.—A system for specifying normal performance characteristics and operating limits of electron tubes by assuming certain usual variations of voltage from various power sources such as a-c power and lighting lines, d-c power lines, and batteries.

detection.—The process in which a high frequency carrier current or carrier wave, modulated with a low frequency signal, is used to produce a new current having a direct component which varies in accordance with the signal. A rectification process applied to high frequency carrier currents. See *detector*. *See illustration*.

detector.—A device furnishing a low frequency or *audio frequency* output which represents the *modulation* of a high frequency or radio frequency current used as the input. A part of a receiver which changes radio frequency power into a form of power suitable for operating a telephone receiver, an audio amplifier, a relay, a telegraphic tape recorder or other form of audible or visible indicator of signals which have been impressed on the radio frequency power. See *plate current detection, grid current detection* and *crystal detector*.

detector probe.—A probe containing a high-frequency rectifying element, either a crystal diode or a tube, which recovers the low-frequency modulation from a carrier and delivers the modulation to a connected instrument such as an oscilloscope, vacuum tube voltmeter, or signal tracer.

detune.—To change either the capacity, the inductance or both in

DEVIATION

a *tuned circuit* so that it no longer is resonant at the applied frequency.

deviation.—In frequency-modulation transmission and reception, variation of frequency above and below the unmodulated or center frequency. Extent of deviation is proportional to audio volume or amplitude.

dial.—A numbered or otherwise graduated disc, segment or drum attached to the shaft of an adjustable device to allow setting of the adjustable quantity to a calibrated value.

diamagnetic.—Tending to lie at right angles with the direction of the lines of force in a magnetic field. Bismuth is the chief diamagnetic material. See *paramagnetic*.

diamond weave coil.—A flat, round inductance coil of the *spiderweb* type in which the turns forming the outer edge take on a diamond shape. *See illustration.*

diaphragm.—The moving member which is acted upon by sound waves in a *microphone*, or which sets up sound waves in a *loud speaker. See illustration.*

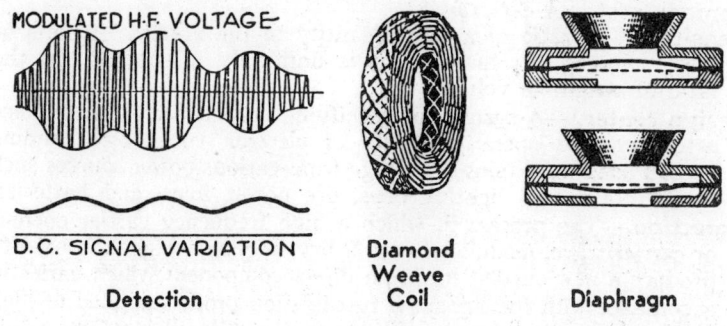

Detection   Diamond Weave Coil   Diaphragm

dichroic mirror.—A surface that reflects light of certain color or wavelength while allowing other colors or wavelengths to pass through.

die-away curve.—A curve passing through the maximum values of a train of *damped oscillations or waves.*

dielectric.—Any non-conducting medium. Generally such a medium when between the plates of a condenser. A material which transmits electric forces by *strain* developed in its mass.

dielectric absorption.—The effect which allows a small current to flow for a short time into a condenser after the first rush of *charging current*, and which causes the instantaneous discharge of the condenser to be followed by a continuously decreasing current.

dielectric conductance.—With a condenser, the conductance of a shunted *equivalent resistance* in which the energy loss is equal to the loss actually due to *dielectric hysteresis*.

DIELECTRIC CONSTANT

dielectric constant.—A characteristic of a substance expressed as the ratio of a condenser's capacity when using the substance as *dielectric* to the capacity of the same condenser when using air as dielectric. The number of times the electrostatic capacity of an air condenser would be increased by using the substance as dielectric between the plates. Also called specific inductive capacity. The usual symbol is K, or the Greek letter epsilon (ϵ).

dielectric current.—*Displacement current.*

dielectric flux.—The electricity which moves in the *dielectric* of a condenser when voltage is applied to the plates. The quantity is equal to that of the charging current. The symbol is the Greek letter psi (ψ).

dielectric flux density.—*Electrostatic flux density.*

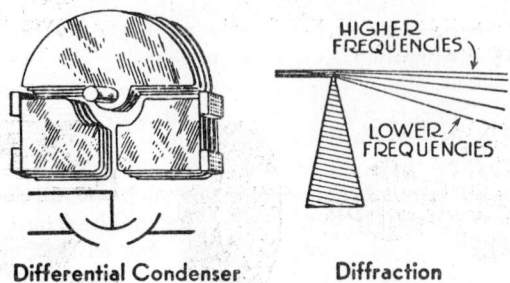

Differential Condenser   Diffraction

dielectric hysteresis.—An effect which retards the rate of charge and of discharge in a condenser, being a sort of electrical friction. The result is that more energy is required to charge the condenser than is secured from it upon discharge, the effect being due entirely to properties of the *dielectric* and not to leakage of current.

dielectric loss.—Dissipation of power in *dielectric conductance* or *dielectric hysteresis.*

dielectric power factor.—The *power factor* of a condenser.

dielectric strain.—The condition of a dielectric substance in which there exists an *electrostatic field* due to charges on the conductors enclosing the dielectric.

dielectric strength.—The greatest voltage per unit of thickness which a *dielectric* will withstand before it breaks down and permits the passage of current. Usually measured in volts per centimeter or per mil (1/1000th inch) of insulator thickness.

dielectric stress.—The number of volts applied per unit of thickness of the *dielectric* in a condenser as the condenser is charged.

difference frequency.—The *beat frequency*, which is equal to the difference between the two frequencies producing it.

differential condenser.—A variable air condenser having two similar sets of *stator plates* and one set of *rotor plates* arranged to mesh first with one and then with the other set of stators. *See illustration.*

differential galvanometer.—A moving coil instrument with two coils, used for comparing two currents or for measuring their difference.

differentiating filter.—One or more capacitors in series with a line and one or more resistors shunted across the line. When capacitive time constant is short compared to the period of an applied voltage in which are rapid changes, the filter output has a series of sharp pulses or pips.

diffraction.—A bending and spreading apart of radio, sound and light waves as they pass around the edges of obstacles in their path. Diffraction increases as the frequency becomes lower. Compare *refraction* and *reflection*. *See illustration, page preceding.*

diffusion.—*Reflection* or *refraction* of waves at various irregular angles rather than in the form of beams.

diode.—A tube in which the only active elements are a cathode and a plate or anode. Primarily a rectifier, allowing electron flow only from cathode to plate. Also a crystal diode.

diode detector.—A diode employed for recovery or separation of low-frequency signals which, at the input, are amplitude modulation on a high-frequency carrier.

dip marker.—An *absorption marker*.

dipole antenna.—Antenna conductors of such overall length that carrier waves of certain frequencies induce simultaneously opposite electric charges or poles at opposite ends. For extraction of signal energy the dipole ordinarily is divided by a center gap at which are connected transmission line conductors.

direct capacitive coupling.—Coupling which results from the reactance of an *electrostatic capacity* which forms a part of each of two coupled circuits. *See illustration.*

direct component.—The average value of a *pulsating current* (or voltage) or of a regularly varying *direct current*, this average value being considered as a steady direct current to which has been added an alternating component to produce the actual pulsating or varying current. Compare *alternating component*.

direct coupled amplifier.—An audio frequency amplifier having a single resistor or capacitor or inductor in the plate circuit of one tube and grid circuit of following tube; bias and plate voltages being secured from a tapped voltage divider. *See illustration.*

direct coupling.—Coupling which is the result of a single inductance, capacity or resistance forming a part of each of the two coupled circuits.

direct current.—An electric current which flows in only one direction through its circuit. A current which is always of the same polarity, although the strength may vary. See *continuous current*. Abbreviated *D.C.*, *d.c.*, or *d-c*.

DIRECT CURRENT AMPLIFIER

direct current amplifier.—An amplifier using *direct coupling* with a resistor included both in the plate circuit of one tube and in the grid circuit of the following tube, suitable grid bias being secured from separate batteries for each stage or by taking the grid connections and plate connections from suitable points on a voltage divider. A change of direct current on the first grid circuit results in a corresponding change of plate current in the last stage.

direct current power supply.—Parts which take power from commercial direct current lines and furnish direct currents at suitable voltages and of suitable freedom from ripple for tube circuits.

direct current resistance.—*Ohmic resistance.*

direct inductive coupling.—Coupling which is the result of the *self-inductance* in a single part which is included in each of the two coupled circuits. *See illustration.*

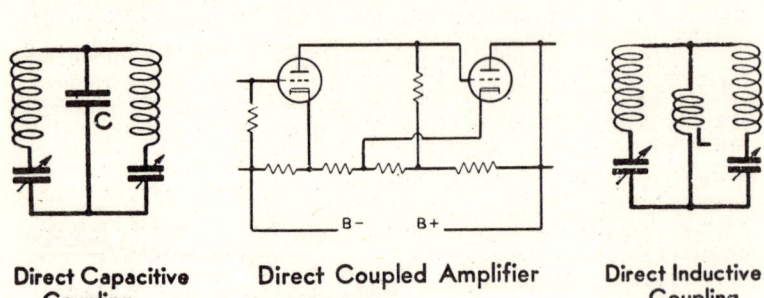

Direct Capacitive Coupling　　**Direct Coupled Amplifier**　　**Direct Inductive Coupling**

direct wave.—1—Any wave which has not been reflected, which comes directly from a source. 2—The *ground wave.*

direction finder.—A radio receiver which allows the direction of travel of radio waves to be determined. A *goniometer* or a *radio compass.*

directional antenna.—An *antenna* which radiates or receives more effectively in some directions than in others.

directional loud speaker.—A *loud speaker* having a projector or radiating surface which delivers sound chiefly over an area included within a narrow angle in front of the loud speaker.

directional response.—The angle or angles within which carrier waves may approach an antenna for maximum signal pickup or for certain fractions of maximum. Usually shown by a polar diagram.

directional selectivity.—The ratio of the force received by an *antenna* from waves coming from a certain direction to the force received from waves of equal intensity coming from all directions.

directive antenna.—An antenna which radiates more energy in some directions than in other directions.

DIRECTLY HEATED CATHODE

directly heated cathode.—A cathode carrying its own heating current and at the same time acting as an electron emitter. A *filament*.

director.—A conductor, or one of several separate conductors, mounted parallel to a dipole antenna on the side approached by received signals. Directors are not conductively connected to the antenna. Part of the wave energy picked up by a director is re-radiated to the dipole, where it reinforces antenna pickup when director spacing and length are suitable for the purpose.

discriminator.—A tube circuit used for demodulation of frequency-modulated signals, also for automatic frequency control. From opposite ends of a transformer tuned secondary are obtained voltages which, while there is no frequency deviation, respectively lag and lead the center frequency by 90 degrees, but which change this phase relation when there is deviation. These voltages combine in two diodes with a voltage from the transformer primary circuit. When there is shift of secondary voltages during frequency deviations the combination voltage increases on one diode and decreases on the other. Diode currents, thus varied by frequency variations, act oppositely in load resistors to produce difference voltages at the audio frequency of modulation.

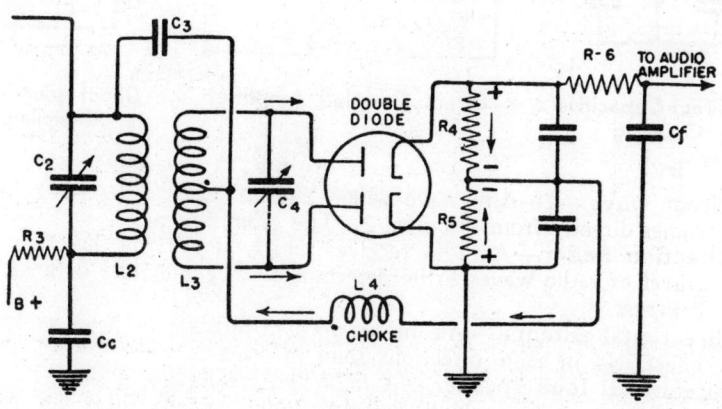

Discriminator

dispersion.—A separation of waves of sound, light or radio energy into beams or rays arranged according to frequency or wavelength, the effect being due to unequal *refraction* of different frequencies.

displacement current.—The momentary flow of current in a dielectric while the *electrostatic field intensity* changes during charge or discharge of a capacity.

disruptive discharge.—Breakdown of a dielectric between bodies having a high potential difference, the breakdown being accompanied by a spark.

dissonance.—1—The relation between circuits not in *resonance* at the same frequency. 2—The relation between wave motions which are not *in phase*.

distortion.—Any change in the form of a wave which occurs during its transmission or amplification. A reproduction of wave form which is not the same in frequency, phase or contour as the original wave. Note that a change in amplitude does not constitute a change in wave form. Compare *frequency distortion, phase distortion* and *amplitude distortion. See illustration.*

distributed capacity.—1—*Electrostatic capacity* existing between extended portions of conductors as distinguished from capacity which is concentrated in condensers. The capacity existing between turns of a winding. *See illustration.* 2—The value of capacity which, connected across an ideal inductance having no distributed capacity, would tune that ideal inductance to the same frequency as that to which an actual coil tunes because of its distributed capacity.

distributed inductance.—1—The inductance which exists along the entire length of a conductor as distinguished from self-induc-

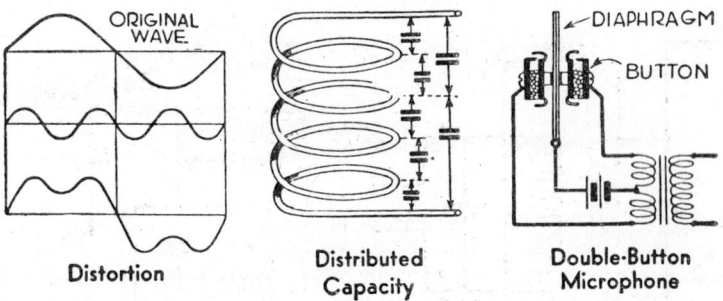

Distortion   Distributed Capacity   Double-Button Microphone

tance concentrated in coils. 2—Inductance added uniformly along a transmission line, to balance the line's capacitive reactance.

distribution center.—The point at which several minor circuits connect to the main supply circuit or to a feeder.

distribution panel.—A switchboard carrying switches and jacks for controlling and interconnecting many circuits.

distribution system.—A system of conductors leading to power consuming devices and carrying current of the form used in such devices.

divergence loss.—The loss in strength of radiated waves due to their spreading out as they travel away from the source. The strength varies inversely as the square of the distance from the source.

donor.—An impurity intentionally added in minute amounts to germanium in forming *n*-type transistor crystals. Donor atoms have more electrons of the kind which may be shared than have germanium atoms, therefore add to the crystal a quantity of electrons which may move freely between atoms.

dope.—A cement used for supporting and insulating the conductors in coils.

dot generator.—A pattern generator producing dots or small rectangles regularly spaced vertically and horizontally on the viewing screen of a television picture tube. Bright dots on a dark background are used during convergence adjustments for color television.

double button microphone.—A *microphone* having two carbon resistance elements or buttons, one on each side of a central diaphragm, and connected in parallel on the current source. *See illustration, page preceding.*

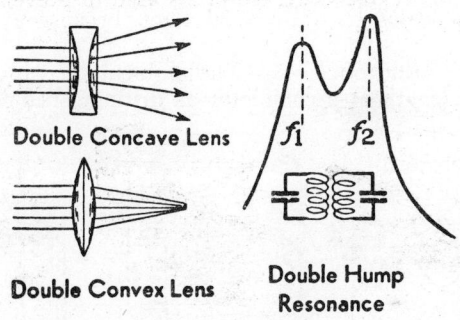

Double Concave Lens

Double Convex Lens

Double Hump Resonance

double concave lens.—A lens of which both sides are concave or recessed. *See illustration.*

double convex lens.—A lens of which both sides are convex or outwardly bulging. *See illustration.*

double conversion.—A system of changing a high-frequency modulated carrier to an intermediate-frequency voltage carrying the same modulation by making the change in two steps. The received high-frequency goes to a superheterodyne oscillator and mixer which make the first reduction in frequency. Mixer output from this first stage goes to a second oscillator and mixer combination in which occurs the final reduction to an intermediate frequency. A radio-frequency amplifier may precede either or both the mixers.

double hump resonance.—The condition with two closely coupled *tuned circuits* wherein *resonance* occurs at two different frequencies, the frequencies being further apart as the coupling is made closer. *See illustration.*

double-magnet ion trap.—An ion trap in which ions and electrons are both turned in the same direction by an electrostatic field between sections of the electron gun. Ions follow this direction and are caught inside the gun, while electrons are turned away from the ion path by one magnetic field, then directed along the axis of the gun and picture tube by a second magnetic field. The magnetic fields are produced by external magnets.

double-pole switch.—A switch which opens or closes two separate circuits or both sides of one circuit.

double refraction.—An effect produced by various crystals which refract or bend a light beam at two different angles, a single entering beam emerging as two beams of which one has a greater angle than the other with the direction of the original beam.

double throw switch.—A switch which connects one conductor to either one of two other conductors. *See illustration.*

doublet.—Equal forces of opposite polarity placed close together.

doublet antenna.—An aerial system including two similar parts placed end to end and having power applied at the center.

draping.—Sound absorbing materials used to prevent excessive *echo* and *reverberation.*

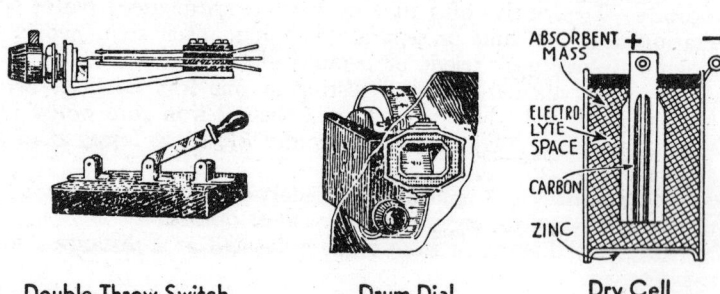

Double Throw Switch   Drum Dial   Dry Cell

dressing.—Positioning and spacing of electrical wiring and parts to avoid undesired couplings and feedbacks.

drive.—Any signal voltage waveform, usually specified by its peak-to-peak value, applied to the grid-cathode input of an amplifier tube or a television picture tube.

driver.—An amplifier capable of furnishing signal power to the grid-cathode input circuit of a following amplifier, usually a push-pull type, in which there is more or less grid current and consequent power dissipation.

drum dial.—A *dial* of which the portion carrying the graduations has the form of a cylinder or drum. *See illustration.*

dry cell.—A *galvanic cell* having a cylinder or box of zinc as the electrode carrying the negative terminal, a carbon rod or block as the electrode carrying the positive terminal and an electrolyte of sal-ammoniac solution with which is mixed zinc chloride. The liquid is held in a mass of manganese dioxide and powdered graphite acting as a *depolarizer.* The voltage is 1.5. *See illustration.*

dry cell battery.—Several *dry cells* connected together in series or parallel to provide more voltage or current than can be taken from a single cell.

dry cell tube.—A tube for which filament current is secured from dry cells.

dry condenser.—An *electrolytic condenser* which is sealed moisture tight.

D.S.C.—Abbreviation for *double silk covered wire*.

dual channel sound.—A television receiver sound system with which signals at the sound intermediate frequency are taken from some point following the mixer output but preceding the video detector, and are amplified at this frequency before demodulation which recovers audio signals.

dual track recording.—Formation of one sound track near one edge and of another track near the opposite edge of a sound recording tape which is run opposite directions through a recording head for producing the two tracks.

dummy antenna.—An *artificial antenna*.

duodiode.—Descriptive of a tube in which are two diode plates operating with the same or separate cathodes, often combined in a single envelope with triode or pentode elements.

dust core.—A magnetic *core* consisting of fine iron particles held by cement. Used where a solid or laminated iron core would introduce too great *eddy current* losses and *hysteresis* losses at high frequencies.

duty-cycle rating.—A continuous or short-time load equivalent to the load imposed during a regular cycle of operation.

DX.—Distant. Reception from stations located at a distance from the receiver.

dynamic.—Relating to motion, and to the forces associated with motion.

dynamic characteristic.—A curve showing the changes produced in one quantity by changes of another quantity under operating conditions. A curve showing the effects of alternating currents. A curve showing the mutual effects of two quantities, both of which vary at the same time. Compare *static characteristic*.

dynamic convergence.—Coming together of the three electron beams in a three-gun color television picture tube at any of the openings which are away from the central area of the aperture plate. Convergence that occurs while the electron beams are being deflected toward the sides, top, and bottom of the viewing screen.

dynamic pickup.—An *electrodynamic pickup*.

dynamic range.—The ratio of maximum to minimum audio signal amplitudes or powers handled without excessive distortion by a sound reproducing system. Usually measured in decibels.

DYNAMOMETER INSTRUMENT

dynamometer instrument.—1—An instrument in which one or more coils move in a field produced by fixed or moving coils. The same current flows through two sets of coils, the two resulting fields reacting upon each other to produce motion of a pointer. The movement used in a *wattmeter*. *See illustration*. 2—An instrument in which a metal disc in the field of a coil is moved by reaction between the coil's field and the field due to eddy currents produced in the disc. The pointer is attached to the disc.

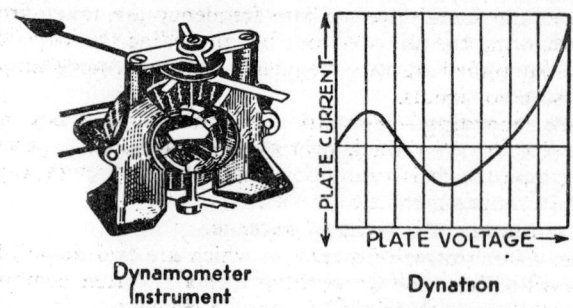

Dynamometer Instrument

Dynatron

dynatron.—A vacuum tube in which *secondary emission* from the plate results in an electron flow to a grid which is at higher potential than the plate. As the plate voltage is increased through a certain limited range, the plate current decreases, exhibiting the effect of *negative resistance*. Generally usd as an oscillator. The actual plate current depends on the difference between the quantity of electrons initially striking the plate and the quantity sent off from the plate by *secondary emission*. *See illustration*.

dyne.—The C.G.S. unit of *force*. The force which would produce a velocity of one centimeter per second when acting on a mass of one gram. Approximately the force exerted by a weight of one milligram acted upon by gravity.

E

E.—Symbol for *effective* voltage, potential difference or electromotive force.

e.—Symbol for *instantaneous* voltage, potential difference or electromotive force.

E_a.—Symbol for *filament* or *heater voltage* at the source.

ear muffs.—Soft rubber rings placed around *headphones* to close the space through which outside sounds might enter the ears.

earphone.—A *headphone*.

earth.—The *ground* or a ground connection.

earth capacity.—A *counterpoise*.

E_b.—Symbol for *plate voltage* at the source.

ebonite.—A name for *hard rubber*.

E_c.—Symbol for *grid bias* voltage at the source. The symbol for supply voltage for the grid nearest the cathode in a multigrid tube is E_{c1}, and for the second grid from the cathode it is E_{c2}.

E. C.—Abbreviation for enamel covered wire.

echo.—A more or less exact repetition of a sound due to the waves being reflected from a surface back toward the source.

echo altimeter.—A *reflection altimeter*.

echo effect.—In sound radio or television reception, the arrival and reproduction at the receiver of two similar signals with a short time interval between them, this effect being due to one of the signals traveling a longer path as it is reflected from the *Heaviside layer* in passing from transmitter to receiver.

echo room.—A room used in broadcasting studios to produce echo effects in order to make a more natural reproduction of certain kinds of sound.

E_d.—Symbol for *screen-grid voltage* at the source.

eddy current.—An electric current which circulates or eddies within the body of a conductor, being produced by varying magnetic fields which pass into the conductor. The fields usually arise from other outside conductors, but may arise from currents flowing in a circuit of which the affected conductor is a part. *See illustration, page following.*

eddy current loss.—The power used in production of *eddy currents*. The heating effect of such currents.

edge effect.—Change in a condenser's apparent capacity due to displacement of the *electrostatic field* at the outer edges of the plates.

edging.—Incorrect or added coloring at the edges or boundaries of objects having different colors in color television pictures.

Edison battery.—An *alkaline storage battery* using powdered iron and mercury in the negative plates, peroxide of nickel and flake nickel for the positive plates, and a water solution of potassium hydrate and lithium hydrate as the electrolyte. The average voltage on discharge is 1.2 volts per cell.

Edison effect.—Flow of current through a vacuum between a positively charged electrode and a second heated electrode. *See illustration.*

editing.—Cutting and removal or parts of a recorded tape and insertion of other parts by splicing during production of a complete desired recording, much as motion picture film is edited.

effective current.—An alternating current value equal to the value of a direct current having the same heating effect. The *root-mean-square* current. For a *sine wave* alternating current the effective value is 0.7071 times the peak value or maximum value. The symbol is I. *See illustration.*

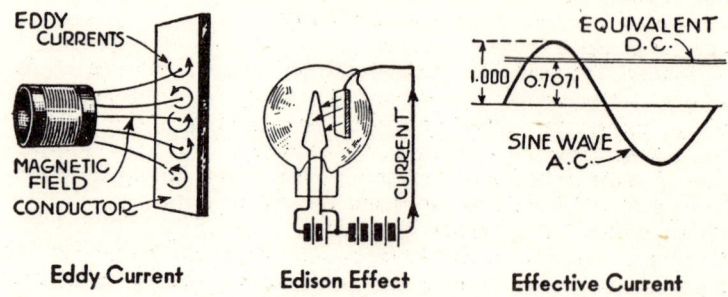

Eddy Current   Edison Effect   Effective Current

effective height of antenna.—The height which would be required in an antenna of similar design, but having no losses, to produce the same radiated field sent out from an actual antenna or to receive from a radio wave the same energy as actually received by the real antenna. An antenna height which is less than the physical height and which may be used in calculations involving *radiation* or absorption of signals. The height of an equivalent antenna consisting of a vertical conductor carrying a uniform current which is the same as the maximum current in the actual antenna.

effective inductance.—*Apparent inductance.*

effective power.—The *true power.*

effective resistance.—1—A measure of the total energy loss in a circuit. Equal to the watts of power divided by the square of the current in amperes in the circuit. 2—In an antenna circuit, the sum of the *radiation resistance,* the *ohmic resistance* and the *dielectric loss.*

effective voltage.—In an alternating current system, the alternating voltage equivalent to a direct voltage which would result in the same heating effect. For a sine wave voltage it is equal to 0.7071 times the maximum or peak voltage. The *root-mean-square* voltage. The symbol is E.

efficiency.—The ratio of the useful energy output to the energy input of a device. Usually expressed as a per cent. The symbol is the Greek letter eta (η).

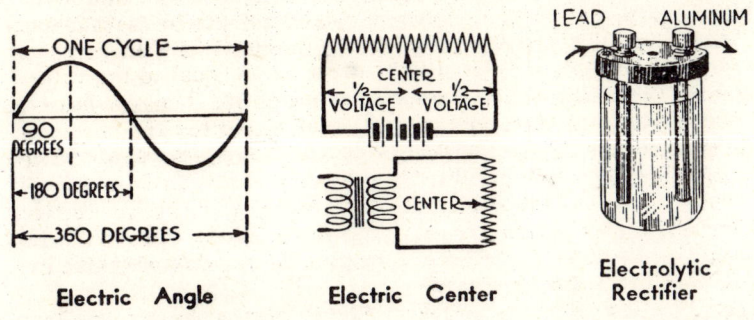

Electric Angle   Electric Center   Electrolytic Rectifier

electric.—1—*Electrostatic*. 2—Pertaining to *electricity* and its effects. Electrical.

electrical.—See under *electric*.

electric angle.—A portion of an alternating current *cycle*. The entire cycle is considered as being divided into 360 equal electric degrees, an electric angle being a part of a cycle measured in degrees and fractions. *See illustration*.

electric attraction.—The force which tends to draw together two bodies carrying *electrostatic charges* of unlike polarity.

electric axis.—An *X-axis* of a quartz crystal.

electric balance.—A *Wheatstone bridge* or an *electrometer*.

electric center.—A point in a circuit or in part of a circuit at which the voltage is midway between the voltages at the ends. A point of zero voltage between other points of positive and negative voltage. A *neutral* point. In a winding carrying alternating current, the point at which there is neither rise nor fall of voltage during a cycle. *See illustration*.

electric charge.—An *electrostatic charge*.

electric circuit.—A conductive path through which may flow *electric current* or between parts of which may exist a *potential difference*.

electric coupling.—*Capacitive coupling*.

electric current.—The flow of electricity in conductors. The rate at which electricity passes a point in a conductor. Measured in *amperes*. The symbol for effective or *root-mean-square* current value is I, and for instantaneous value it is i.

electric degree.—One of the 360 equal parts of a *cycle*. See *electric angle*.

electric density.—The amount of electricity existing as a *charge* per unit of area on a conductor's surface.

electric displacement.—Movement of electrons with reference to the molecules to which they are bound in a *dielectric*. *Displacement current*.

electric field.—An *electrostatic field*.

electric fluid.—An obsolete term for electricity.

electric pole.—A point or a surface at which is exhibited an *electrostatic force;* a surface at which an *electrostatic field* enters or leaves a medium.

electric repulsion.—The force which repels from each other two bodies carrying *electrostatic charges* of like polarity.

electrification.—The condition of a body as determined by the quantity of *electrons* it carries. A body is positively electrified when it carries less than a normal number of negative electrons (resulting in a positive charge) and is negatively electrified when carrying more than the normal number of negative electrons (resulting in a negative charge).

electrify.—To produce an *electrostatic charge* on an insulated conductor.

electro-acoustic.—Relating to the effects of electric currents and sound recording or reproducing devices on each other.

electro-acoustic transducer.—Apparatus transferring power in either direction between an electric circuit and an *acoustic* device.

electrochemical.—Relating to *electrochemistry*. Electrolytic.

electrochemistry.—The science of the production of chemical effects from electricity and of the production of electric energy from chemical energy.

electrode.—A terminal or surface at which electricity passes from one material or medium into another. An *anode* or a *cathode*. A conductor by which current enters or leaves any electrical device.

electrode conductance.—The ratio of current change in the circuit of a tube electrode to the change in voltage on the same electrode, other potentials remaining constant.

electrodynamic.—Relating to the effects between electric currents and the forces they produce through their *magnetic fields*. The science of electrodynamics deals with the action of electric currents on themselves and on other currents, and with actions between currents and magnets.

electrodynamic induction.—*Self-induction* or *mutual induction*.

electrodynamic loud speaker.—A *moving coil loud speaker*.

electrodynamic pickup.—A phonograph pickup with which movement of the needle in the gap of a permanent magnet varies the magnetic flux to induce voltage and current in a coil around the magnet.

electrodynamometer.—A *dynamometer*.

electrokinetic energy.—Energy contained in electricity in motion; that of the *electric current* and its *magnetic field*.

electrokinetics.—The science of electricity in motion. Compare *electrostatics*.

electrolysis.—Chemical change or decomposition caused by flow of electric current. Separation of a liquid compound into its elements by an electric current flowing through it, the elements being deposited upon an electrode in the process or liberated as gases.

electrolyte.—A liquid chemical compound in which flow of electric current causes separation of the elements of the liquid, or a chemical change in the liquid. The liquid used in *primary cells, storage cells,* or *electrolytic cells*.

electrolytic capacitor.—A dry electrolytic capacitor, the kind commonly used, has plates of aluminum or other metal foil with plain or variously formed surfaces, separated by some porous material carrying liquid electrolyte which is the negative electrode and with which contact is made through a cathode foil. The dielectric is a very thin layer of oxide on the anode foil. Capacitance is great in relation to capacitor bulk because of the exceedingly thin dielectric. Common types have high resistance to current flow in one direction but relatively low resistance in the opposite direction, and must be connected with due regard to polarity of direct voltages in a circuit.

electrolytic rectifier.—One or more plates of lead, iron, carbon, or other inert material, and one or more plates of aluminum or tantalum immersed in an electrolyte solution. Electric current will pass from the electrolyte into the aluminum or tantalum, but not in the reverse direction, thus allowing *rectification* of alternating current. Current tending to flow from the aluminum or tantalum into the electrolyte causes the immediate formation of a very thin film of insulating gas on the metal's surface, thus stopping the current flow. Reversal of voltage causes instant decomposition of the gas film and current is allowed to flow. *See illustration.*

electrolyze.—To separate by *electrolysis*.

electromagnet.—1—A soft iron *core* wholly or partially surrounded by a coiled conductor, flow of current through the conductor making the iron a magnet as long as the current continues to flow. *See illustration.* 2—An *air-core* winding which exhibits all the attributes of a magnet while current flows in the winding.

electromagnetic.—Pertaining to or depending upon *electromagnetism*.

electromagnetic component.—That portion of *radiation* which is due to electromagnetic fields producing detached magnetic loops and consequent disturbances propagated through space.

electromagnetic coupling.—*Inductive coupling*.

electromagnetic convergence.—Magnetic convergence.

electromagnetic deflection.—Magnetic deflection.

ELECTROMAGNETIC FIELD

electromagnetic field.—The lines of force and the space they occupy around an electromagnet. A *magnetic field*.

electromagnetic flux.—The flow of magnetism through an electromagnet and its field; the *lines of force* passing through the magnet and its field. *Magnetic flux*.

electromagnetic focus.—Magnetic focus.

electromagnetic induction.—The production of *electromotive force* in a conductor which is cutting through magnetic lines of force or which is being cut through by moving lines of force. The principle employed in the electric generator and in inductive coupling. *Self-induction* and *mutual induction*. See illustration.

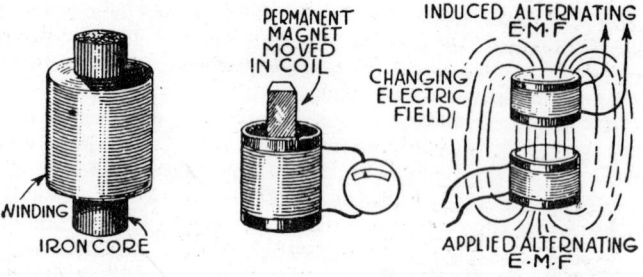

Electromagnet   Electromagnetic Induction

electromagnetic microphone.—A *microphone* in which sound waves cause vibration of a coil in a strong magnetic field, currents at audio frequency being produced by voltages generated in the coil.

electromagnetic pickup.—A *magnetic pickup*.

electromagnetic radiation.—Sending into space the lines composing an electromagnetic field; one portion of the total *radiation* from a transmitter.

electromagnetic rectifier.—A *vibrating rectifier*.

electromagnetic stress.—*Magnetic stress*.

electromagnetic unit.—Any *absolute unit* in the C.G.S. system; a unit based on the force exerted between two magnetic poles.

electromagnetic wave.—A periodic disturbance arising from electromagnetic action and producing a varying magnetic field and a varying electric field traveling through space. A radio wave.

electromagnetism.—1—*Magnetism* resulting from flow of current. 2—The relations between *magnetism* and *electricity*.

electrometallurgy.—The use of electricity in producing metals.

electrometer.—1—An instrument for measuring *potential differences*. In one type the potential produces electrostatic charges on two stationary plates, one of which is connected to a movable vane of which the deflection is proportional to the square of the effective applied voltage. 2—An *electroscope*.

electromotive force.—The *force* which acts to move electricity in the form of the electric current. Measured in *volts*. (*e.m.f.*)

electron.—The smallest quantity of negative electricity which is capable of moving between atoms of matter. A unit negative charge. One of the particles of negative electricity which is emitted by the *cathode* in a vacuum tube and is attracted toward a positively charged *anode* in the tube.

electron beam.—Electrons emitted from the cathode of a television picture tube or cathode-ray tube, focused into a stream of small diameter, and drawn by high potential to the viewing screen where they excite the phosphors.

electron bombardment.—Striking of *electrons,* moving at a high velocity, against the surface of a conductor. This bombardment may result in liberation of other electrons from the surface, in *secondary emission*.

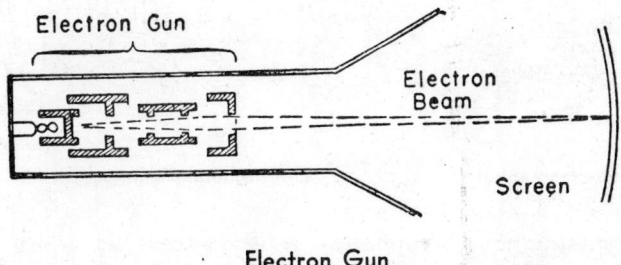

Electron Gun

electron coupling.—Varying the electron flow from cathode to plate in a tube by action of elements which do not include the plate and, therefore, are not in the output circuit. Coupling between active elements and the output circuit is by means of the single common electron stream. For example, the control and screen grids may be in an oscillator circuit, the suppressor an isolating element, with the plate delivering oscillating current to connected circuits.

electron emission.—1—A passage of *electrons* from the surface of a material into the surrounding space due to the action of heat, light, cathode rays, chemical action or impact excitation. 2—The rate of electron emission from a cathode, measured as the current resulting when all emitted electrons are drawn from the cathode.

electron flow.—Movement of free electrons from negative to positive points in metals or other conductors, or from negative to positive electrodes through liquids, gases, or vacuums.

electron gun.—In a television picture tube or cathode-ray tube, an assembly including cathode, first and second grids, all or part of the anode, and sometimes a focusing electrode. This assembly furnishes the electron beam.

ELECTRON THEORY OF MAGNETISM

electron theory of magnetism.—The theory which holds that the rotation of *electrons* in the atoms of a magnetized body may be changed so that most of them act to form magnetic fields acting in one direction. It is assumed that rotation of an electron produces a magnetic effect just as rotation of electric current in a coil produces such an effect.

electron theory of matter.—The theory that the *electron* is a common part of all substances, that the *atoms* of all elements are composed of positively charged central parts about which rotate the negatively charged electrons in definite orbits or paths. In the space between atoms it is assumed that there are *free electrons* which are moved by application of *electromotive force* when the material considered is a conductor.

electron tube.—A *vacuum tube* utilizing the effect of emission of electrons from a heated body within the tube. The type of tube commonly used in radio applications.

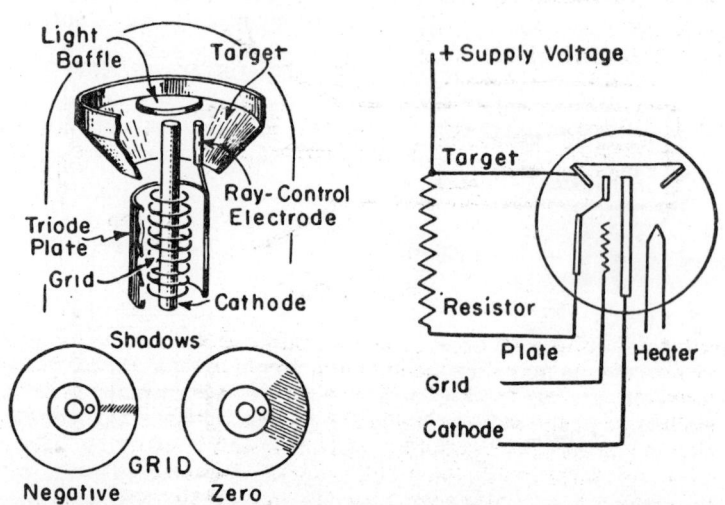

Electron-ray Tube

electronic switch.—An instrument which allows simultaneous display of traces for two separate voltages on the screen or an ordinary oscilloscope designed for display of only a single trace.

electronics.—The science dealing with the action and effects of electron flow in vacuums and gases, and with the use of vacuum tubes and gaseous tubes in all branches of art and industry.

electron-ray tube.—A vacuum tube within which is a plainly visible target surface coated with fluorescent material which glows when struck by a beam or ray of electrons from a cathode extension. The size of the ray and of the glow change with variation of negative charge on a small ray-control electrode between cathode and target.

electrostatic.—Relating to *electrostatic charges* or to electricity at rest. The effects observed in capacities, in condensers and in the fields or lines of force between charged bodies are electrostatic in their nature. The science of electricity at rest, and of electric charges, is called electrostatics. Compare *electrokinetics*.

electrostatic capacity.—The ability to hold an *electrostatic charge*. The ratio of the number of *coulombs* of electricity stored in a condenser to the voltage difference between the plates. Also called capacitance and permittance. Measured in *farads,* microfarads, etc. The symbol is C.

electrostatic charge.—The amount of electricity, measured in *coulombs*, which is contained in a condenser and which will flow through a circuit connected to the condenser terminals. The amount of electricity caused to flow from a condenser by release of the dielectric strain. A similar amount of electricity contained in or delivered from any capacity.

electrostatic circuit.—The path in which exist *electrostatic lines of force* in dielectric materials.

electrostatic component.—The portion of *radiation* which is due to electrostatic fields. See *electromagnetic component*.

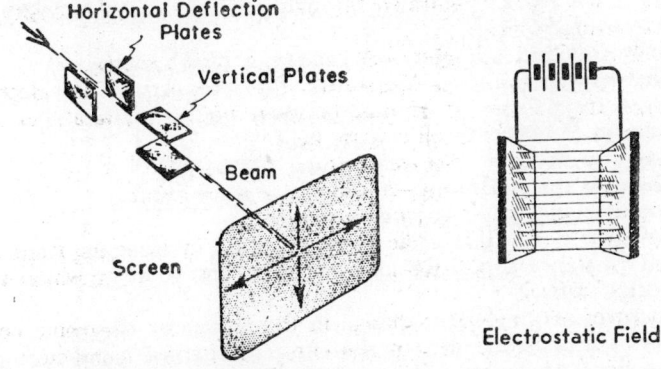

Electrotstatic Deflection   Electrostatic Field

electrostatic convergence.—Convergence in a three-gun color television picture tube by means of potentials applied to convergence electrodes within the tube.

electrostatic deflection.—Vertical and horizontal bending of the electron beam in a cathode-ray tube by means of electric fields formed between a focusing electrode and other electrodes through which the beam passes.

electrostatic energy.—Energy contained in electricity at rest, such as in the charge of a condenser. Electrical *potential energy*.

electrostatic field.—The space traversed or occupied by *electrostatic lines of force*. A space within which forces exist because of electrostatic charges on conductors on either side of the space. *See illustration.*

electrostatic field intensity.—1—The force exerted on a *unit pole* of positive polarity at a point in an electrostatic field. 2—*Radio field intensity.* 3—The ratio of the voltage across a capacity to the distance between the plates, usually measured in millivolts or microvolts per meter or per centimeter.

electrostatic flux.—The *electrostatic lines of force* existing between bodies at different potentials.

electrostatic flux density.—The number of electrostatic lines of force per square centimeter cross section of an *electrostatic field*. The symbol is D.

electrostatic force.—Force exerted by attraction and repulsion between *charged bodies*, or bodies between which there is a difference of potential. Electric attraction and electric repulsion.

electrostatic induction.—The production of *electrostatic charges* on a conductor which is brought into an *electrostatic field*. Any insulated conductor brought near another conductor which is charged will receive charges by means of electrostatic induction.

electrostatic leakage.—*Condenser leakage.*

electrostatic lines of force.—Lines indicating the paths in which are acting the forces of attraction or repulsion between electrically *charged bodies*. Lines drawn through points of equal intensity in an *electrostatic field*.

electrostatic loud speaker.—A *condenser loud speaker.*

electrostatic machine.—Apparatus for production of electric charges by means of friction between unlike materials or by movement of bodies in an electric field.

electrostatic meter.—See *electrostatic voltmeter*.

electrostatic microphone.—A *condenser microphone.*

electrostatic pickup.—A *condenser pickup.*

electrostatic potential.—The work required in bringing from an infinite distance a positive *unit charge* to the point at which the potential exists.

electrostatic strain.—The change in the atomic or electronic condition of a *dielectric* due to the effect of charged conductors on either side. Analogous to a mechanical strain which accompanies a change in shape, form or size of a body subjected to a force.

electrostatic stress.—The force exerted by charged bodies upon a *dielectric* between them. Measured in volts per unit thickness of the dielectric. The symbol is G.

electrostatic unit.—An *absolute unit* in the C.G.S. system, a unit based on the force between two quantities of electricity or on the properties of a *unit charge* of electricity.

electrostatic voltmeter.—An instrument measuring potential differences by the repulsion or attraction between conductors which are charged by the voltage being measured. These meters may be used with either D.C. or A.C. voltages, and they take negligible current from the circuit. *See illustration.*

electrostatics.—The science of electricity at rest, and of electric charges. Compare *electrokinetics.*

electrostriction.—A change in the physical form or size of a body which is being acted upon by *electrostatic stress* or by an electric field.

electro-therapeutic.—Pertaining to the use of electricity in medicine and surgery.

electrothermic.—*Thermoelectric.*

element.—1—One of the *electrodes* in a tube, a cell, or other electrical device. 2—A substance which cannot be separated by ordinary physical or chemical means into other substances.

e.m.f.—An abbreviation for *electromotive force.*

emission.—*Electron emission.*

emission characteristic.—A graph showing the relation between the *total emission* from a cathode and the effect controlling the emission; a voltage, current, temperature, etc.

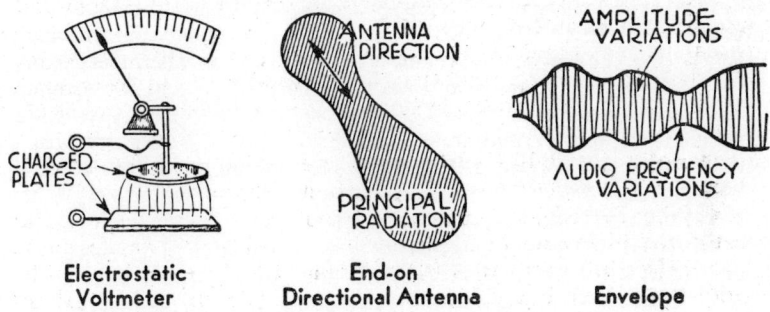

Electrostatic Voltmeter

End-on Directional Antenna

Envelope

emitter.—One of the elements in a transistor. Electrons pass from an n-type base to the emitter, or from the emitter to a p-type base. The emitter portion of the crystal is of type opposite to that of the base in junction transistors, or is a small wire in point-contact transistors.

empire cloth.—An insulating fabric made from cambric impregnated with oils.

empirical.—Based on experience and observation but not always on recognized physical laws.

end-on directional antenna.—A directional antenna radiating chiefly in the direction of the line on which the antenna elements are disposed. Compare *broadside directional antenna. See illustration.*

energy.—The ability to do *work*. An ability due to motion, position, electromotive force, chemical action, temperature, etc. Electrical energy is measured in *joules* and in *watt-hours*. The symbol is W. See *potential energy* and *kinetic energy*.

envelope.—1—The evacuated glass bulb or metal shell of an electronic tube within which are the active elements. 2—Lines or curves drawn through maximum amplitudes of an alternating quantity to show variations with time, or to show waveform of voltage or current obtained upon demodulation.

E_p or e_p.—Symbol for *plate voltage* between plate and cathode.

equalizer.—In an audio-frequency amplification system, circuits or filters designed to compensate for variations in frequency response of other portions of the system, or of input signals.

equalizing pulses.—In the composite television signal, six pulses that precede and another six that follow each complete serrated vertical pulse, being of the same amplitude as sync pulses and of half the duration of horizontal sync pulses. The purpose is to allow alternate picture fields to begin with a full line and with a half line for interlacing.

equivalent circuit.—An arrangement of series and shunt connected impedances which is the equivalent in action of a complicated circuit at a given frequency.

equivalent focus and focal length.—The *focus* or the *focal length* which actually results from the use of several lenses in one system, or such a value as translated into terms of a single lens producing the same effect.

equivalent periodic line.—A *periodic line* having the same electrical behavior as a *smooth line* at some assumed frequency.

equivalent resistance.—The amount of *resistance* which would cause the total amount of energy loss which actually takes place in an electrical part. It is the resistance which would have to be added to a loss-less circuit of the same type in order that an equal loss might take place. Equivalent series resistance is the resistance which would be added in series, and equivalent shunt resistance is the resistance which would be added in shunt with the ideal circuit or device. In a condenser it is the equivalent series resistance representing the loss due to *dielectric absorption*. *See illustration.*

equivalent sectional radius.—The square root of the cross section normal to the axis, divided by pi (π).

equivalent sine wave.—A *sine wave* which has the same effective value and the same frequency as an actual wave of alternating current.

equivalent smooth line.—A *smooth line* having the same electrical behavior as a *periodic line* at one frequency.

erasing head.—On a tape or wire recorder or reproducer, a member in which is an electromagnet for removing the magnetization which represents a recording.

erg.—The C.G.S. unit of *work* or *energy*. The work done by one dyne of force acting through one centimeter distance. The energy used in moving a body one centimeter against a force of one dyne. 10,000,000 ergs equal one *joule*.

E$_s$.—Symbol for *screen voltage* between screen and cathode in a vacuum tube.

eta (η).—Greek letter symbol for *efficiency*.

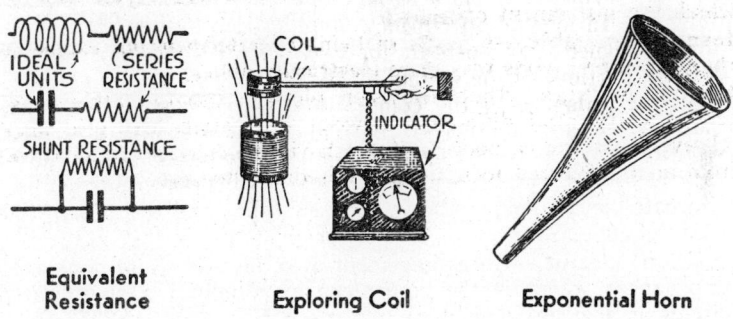

Equivalent Resistance   Exploring Coil   Exponential Horn

ether.—A medium which has been assumed to pervade all space regardless of what objects or substances are occupying that space. Radio waves, light, heat, X-rays, etc., are assumed to cause movements of the ether and to be propagated through it.

ether waves.—Radio waves, light waves, radiant heat waves, etc., which are transmitted through and by the ether.

evacuation.—The process of withdrawing air and other gases from within the bulb of a tube, leaving a more or less perfect vacuum.

excitation.—Production of electrical effects.

expansion type meter.—A *hot wire meter* in which lengthening of the current carrying conductor allows movement of a pointer.

exploring coil.—A small *inductance coil* connected to some indicator of current flow, such as a detector and headphone. When the coil is moved into a changing field, currents are produced in its circuit and are detected or indicated to allow determination of the field's extent and intensity. *See illustration.*

exploring disc.—A *Nipkow disc*.

explosion-proof.—Descriptive of apparatus in a case constructed to withstand an explosion of specified gas or dust within it, and prevent ignition of the gas or dust around the case by sparks or flashes.

exponential damping.—*Damping* which proceeds according to an exponential law.

exponential horn.—A sound projector or horn of which the cross sectional area increases with its length according to a logarithmic law, successive areas at equal distances from each other increasing at a constant ratio from one to the next. *See illustration.*

exponential notation.—See *standard notation.*

exponential tube.—A form of *variable-mu tube.*

extensional vibration.—*Longitudinal vibration.*

external field influence.—The percentage change caused in an instrument's reading by an external field having an intensity of five *gausses* in the most unfavorable phase and position, the field being produced by a current of the same frequency as that at which the instrument operates.

externally operable.—Capable of being operated without exposing the operator to parts which are electrically alive.

extinction voltage.—In a television picture tube or cathode-ray tube, the potential difference between a negative first grid and relatively positive cathode with which viewing screen illumination from an undeflected focused spot is reduced to zero.

F

F.—Symbol for *magnetomotive force*.
F. or f.—Symbol for *luminous flux*.
f.—Symbol for *frequency* in cycles.
face-parallel cut.—A *Y-cut* for a quartz crystal.
face-perpendicular cut.—An *X-cut* for a quartz crystal.
face plate.—The glass front of a television picture tube, inside of which is the viewing screen with phosphor coating, or a phosphor plate.
facsimile transmission.—Electrical transmission of any representation which may be made in black and white or with very limited changes in shade, such as writing, diagrams, etc. *Telephotography*.
fader.—A *mixer*. A volume control for an audio frequency amplifier.
fading.—A periodic rise and fall of strength of radio wave signals arriving at a receiver; assumed as being due to interference between waves reflected at various angles from the *Heaviside layer*. Variation of signal strength due to changes occurring in the transmission path.
Fahnestock clip.—A form of binding post having a spring catch which holds and makes contact with a wire.
Fahrenheit temperature scale.—One of the scales used for measurement of temperature. The temperature of melting ice is at 32 degrees above zero and the temperature of boiling water is 212 degrees above the zero point. Compare *centigrade temperature scale*.
falling characteristic.—The result of *negative resistance*, especially in an arc where increase of current causes greater ionization and allows a still greater current, the equivalent of a decreasing resistance with an increase of current. *See illustration*.
fan antenna.—A dipole antenna whose signal pickup conductors on opposite sides of a central support diverge at small angles to one another. A name sometimes applied to a conical antenna or to a bow tie antenna.
farad.—The practical unit of *electrostatic capacity*. A capacity with which a potential difference of one volt between plates results in a charge of one coulomb of electricity, or in a charging current of one ampere flowing into the capacity for one second.
Faraday effect.—Rotation of a beam of polarized light passing through a magnetic field.
Faraday's dark space.—A non-luminous region near the cathode, and between the visible glows around the cathode and anode, in a tube containing gas undergoing ionization.

Faraday's law.—The law which states that the induced *electromotive force* in a circuit is proportional to the rate of change of magnetic lines of force linked with the circuit.

feedback.—Transfer of signal energy from output to input of the same system. If feedback voltage or current is in phase with voltage or current already existing at the input there is increase of total input and the effect is called regeneration. Feedback in opposite phase reduces total input and causes degeneration.

fibre.—An insulating and supporting material made chiefly from treated paper, cloth and cellulose. Dielectric constant 5.0 to 8.0.

fidelity.—The degree in which an amplifying system or other circuit delivers from its output an accurate reproduction of the input signal. The opposite of *distortion*.

field.—1—A space within which are magnetic forces from a magnet or a current-carrying conductor, or electrostatic forces from electric charges. 2—In television pictures, all the alternate horizontal lines formed during one downward travel of the electron beam. A picture is completed by tracing the intervening lines during the following downward travel of the beam, the method being called interlacing. Two successive fields, taking 1/60 second each, form one frame whose period is 1/30 second.

field frequency.—A television signal frequency of 60 cycles per second, the frequency at which successive fields occur.

field intensity.—*Magnetic field intensity, electrostatic field intensity* or *radio field intensity*.

field magnet.—An electromagnet or a permanent magnet furnishing a strong magnetic field in a loud speaker, a microphone, a phonograph pickup or other electrical device.

field neutralizing coil.—A few turns of conductor around the outer edge of the face of a color television picture tube. Current in the coil produces a magnetic field adjusted to counteract various stray fields in the vicinity of the tube.

field strength.—Strength of carrier signals at a given locality, as measured in the number of microvolts of signal obtained from an antenna. Absolute field strength, measured in microvolts per meter, is the signal voltage obtained from a standard antenna one meter long from end to end.

filament.—A tube cathode which is heated by current flowing through it and from which electron emission takes place.

filament battery.—A battery which provides current for heating a tube filament.

filament capacitance.—The sum of the separate *direct capacitances* between a tube filament and all the other elements in the tube. In a three-element tube, the sum of the *grid-filament capacitance* and the *plate-filament capacitance*.

filament cathode.—An electronic tube cathode which carries heating current while emitting electrons which flow to an anode or other element.

filament circuit.—All of the parts through which flows current for heating a tube filament. *See illustration.*

filament current.—The electric current used for heating the filament in a tube; the current flowing through the filament. The symbol is I_f.

filament resistance.—The *ohmic resistance* of a tube filament. The symbol is R_f.

filament resistor.—A resistor, usually of the fixed type, which limits the flow of current through a tube filament.

filament voltage.—The *potential drop* across the filament of a tube. The voltage applied across the ends of a filament. The symbol for voltage of the source is E_a, and for voltage across the filament is E_f.

filter.—1—An electric filter is a combination of resistances, induc-

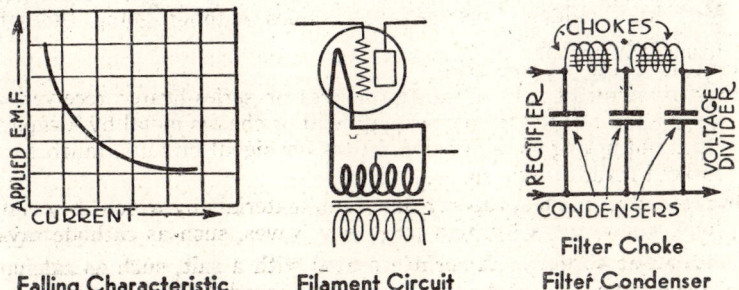

Falling Characteristic   Filament Circuit   Filter Choke / Filter Condenser

tances and capacities, or any one or two of these; used to attenuate currents and power at certain frequencies while allowing comparatively free flow of other frequencies or of direct current. These filters are classed as *high pass, low pass, band pass* and *band exclusion* filters. 2—Light filters and sound filters are devices which attenuate or completely prevent the passage of light or sound of certain frequencies or wavelengths.

filter choke.—An inductance used in a *filter* system to retard the flow of currents at higher frequencies and permit passage of low frequencies or direct current. *See illustration.*

filter condenser.—A capacity used in a *filter* system to prevent the flow of direct current or retard flow of currents at low frequencies and permit passage of higher frequencies. *See illustration.*

fine tuning.—A television receiver adjustment, usually a variable capacitor, for making small changes of resonant frequency in the radio-frequency oscillator of the tuner.

first detector.—In *superheterodyne reception,* the tube in whose grid circuit the signal frequency and the oscillator frequency combine to produce the *intermediate frequency* or the beat frequency which appears in the plate circuit.

fix.—The point at which lines for two or more *radio bearings* cross one another.

fixed capacitor.—Any capacitance whose capacitance is not adjustable or variable, but is fixed during manufacture.

fixed resistor.—A resistor whose resistance is not adjustable but is fixed during manufacture.

flexible armored cable.—A *cable* covered with two layers of spirally wound metal strip.

flexible cord.—A *stranded conductor* within a flexible insulating covering.

flexible metal conduit.—A metal tubing composed of spirally wound strips allowing easy bending. Used for enclosing and supporting all classes of wiring.

flexible tubing.—*Circular loom. Spaghetti tubing.*

flip-flop.—Descriptive of an oscillator whose operating frequencies alternates suddenly between two values without going through intervening frequencies.

floating grid.—A free grid.

floating ground.—In a transformerless or series-heater receiver, a conductor isolated from chassis ground or chassis metal by a capacitor, and acting as a common return for signal circuits, much as a ground return might act.

fluorescence.—The property of certain materials by which they emit light when struck by high frequency waves, such as cathode rays.

fluorescent screen.—A surface coated with a salt, such as calcium sulfide, which emits light when bombarded by electrons, as cathode rays, X-rays, etc. The large end of a cathode-ray tube.

fluoroscope.—A device making use of X-rays to produce *fluorescence* shadows of objects surrounded by a material through which visible light does not pass.

flutter.—Rapid changes of sound frequency from a tape, wire, or phonograph reproducing system, due to variations of speed in either the recording or reproducing apparatus.

flutter echo.—An *echo* consisting of numerous distinct repetitions of the original sound.

flux.—The passage of light, sound, heat, magnetism, radio waves or other forms of *radiant energy* through a space. The rate at which such a flow takes place in a given cross section of the space. The lines of force which constitute a *magnetic,* an *electrostatic* or a *radiation field.*

flux density.—The number of lines of force passing through a given cross sectional area in a field. See *magnetic flux density* and *electrostatic flux density.*

flux linkage.—The product of the number of turns in a coil and the number of *magnetic lines of force* passing through the coil.

flux meter.—An instrument for measuring and indicating *magnetic flux* in a field.

flyback period.—The time during which a television scanning beam is extinguished while voltages which move it return to values at which the following trace will start to cover the picture area.

flyback power supply.—A high-voltage supply system for anodes of television picture tubes. Voltage pulses induced in horizontal output transformer windings by reversals of deflecting current during retrace or flyback are increased in voltage by an autotransformer extension of a winding, then rectified and filtered.

flywheel effect.—Continued oscillations at the resonance frequency in a tuned circuit after the exciting voltage is no longer applied. Amplitude decreases rather slowly when the circuit is of high-Q construction.

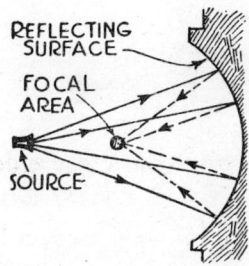

Focal Area for Sound

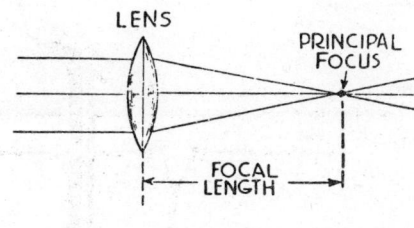

Focal Length

focal area for sound.—The concentration of sound waves reflected from a curved surface. *See illustration.*

focal length.—The distance between the *optical center* and the *principal focus* of a lens. *See illustration.*

focal plane.—A plane lying parallel to the face of a lens, at right angles to the principal axis, and at such a distance from the lens that it is the location of the principal focus formed by the lens.

focus.—A point at which reflected or refracted waves or rays come together. In optics, a *real focus*, a *principal focus*, a *virtual focus*.

focusing.—In a television picture tube or cathode-ray tube, concentration of the electron beam into a small spot at the point where it strikes the viewing screen. Magnetic focusing or electrostatic focusing.

folded dipole.—A dipole antenna whose signal-pickup element consists of two or more horizontal conductors lying parallel to one another and joined at their outer ends. Signal takeoff is at a gap in one of the conductors.

folded picture.—Television picture distortion resulting from lack of synchronization of electron beam deflection with picture-producing signals. Parts of the picture appear to overlap, either horizontally or vertically.

foot-lambert.—A unit of *brightness*. The brightness of a surface which is emitting or reflecting one *lumen* per square foot.

force.—Any effect which changes or tends to change the condition of objects; any agency which tends to change the position of an object or to change its ability to do work. Measured in *dynes*.

force factor.—A measure of the coupling between the electrical and mechanical systems of an *electro-acoustic transducer*.

forced oscillations.—Oscillations which occur in a circuit as the result of an impressed voltage having the frequency of the oscillations, often different from the circuit's *natural frequency*.

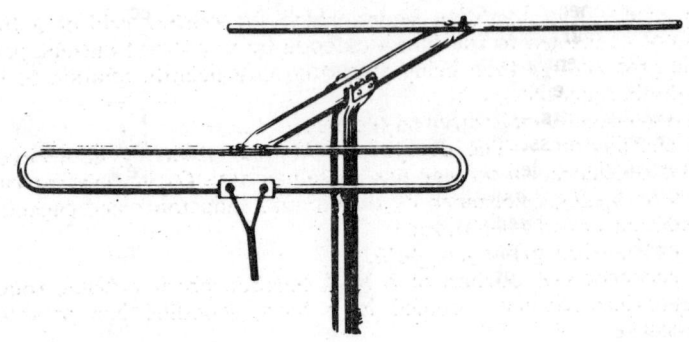

Folded Dipole

form factor.—The ratio of the effective value to the mean value in one *alternation* of a wave. The form factor of a sine wave is 1.111.

formica.—The trade name of an insulating *phenolic compound*.

form winding.—Winding and shaping a coil upon a fixed form before mounting in its working position.

Foster bridge.—A *slide wire bridge* for comparing two resistances.

Foucault current.—An *eddy current*.

four-element transistor.—A junction transistor whose crystal consists of alternate n-type and p-type sections. Three of the sections act as base, emitter, and collector. The fourth, which may be of either n-type or p-type, may be employed in various ways to alter the gain, the frequency response, or other properties of the circuit.

four-terminal transistor.—A junction transistor having one terminal for emitter, another for collector, and two for the base. The additional base terminal may be biased to lessen effective resistance of the transistor.

frame.—In television pictures, all of the horizontal lines which complete the images during two fields or two successive downward travels of the electron beam for interlaced scanning. One frame consists of two successive fields.

frame frequency.—A television signal frequency of 30 cycles per second, 1/30 second being the period for one complete frame.

free alternations.—Alternations having a frequency determined by the *natural frequency* of the circuit in which they occur.

free charge.—1—Instantaneous current upon discharge of a condenser. 2—A charge not bound. See *bound charge*.

free electron.—An *electron* which does not move in a regular orbit around the positive nucleus of an atom but follows an irregular path, leaving the atom entirely upon formation of the electric current.

free grid.—The condition under which the *control grid* of a tube is not connected to the tube's cathode by any direct current path, the *grid voltage* then being free from any definite relation to the cathode potential.

free impedance.—*Iterative impedance*.

free oscillations.—The *damped oscillations* occurring in a circuit after an impressed voltage has ceased to act. *Oscillatory currents* at a frequency determined by the circuit's inductance and capacity. Compare *forced oscillations*.

free path.—The *mean free path*.

free radiator.—A portion of a loud speaker which radiates sound waves into air not confined in a horn, sounding box or other enclosure.

free running frequency.—An oscillator frequency determined solely by values of capacitance and of inductance or resistance in its circuit. The frequency when operation is not controlled by any applied synchronizing voltage.

frequency.—The number of complete *cycles* per second existing in any form of wave motion; as the number of cycles per second of an alternating current, a sound wave or a light beam. Measured in *cycles*. The symbol is f.

frequency band.—All the frequencies between two definite limiting frequencies.

frequency compensation.—Any of various methods which extend the range of frequencies within which an amplifier has nearly uniform gain. The frequency may be extended higher or lower, or both ways. High-frequency compensation or low-frequency compensation.

frequency correction.—Compensation for unequal transmission of various frequencies in a line by use of *attenuation equalizers*.

frequency discrimination.—*Frequency distortion*.

frequency distortion.—A type of distortion in which certain *frequencies* are amplified or attenuated more or less than are other frequencies.

frequency doubler.—Apparatus for doubling an alternating current frequency; generally by selection and amplification of the second *harmonic* of the original frequency.

frequency drift.—A change of oscillator frequency during operation, usually due to variations of capacitance or inductance brought about by heating of circuit elements.

frequency influence.—The percentage change in an instrument's reading caused by a deviation of ten per cent from the rated frequency.

frequency meter.—A device which allows measurement of the *frequency* of an alternating current or voltage, or the frequency of a radio carrier wave. The usual form consists of a tuned circuit calibrated in frequencies and provided with means for showing maximum current or other indication of resonance when the tuned frequency of the circuit is made the same as the frequency of the voltage, current or field to be measured. Compare *autodyne frequency meter*.

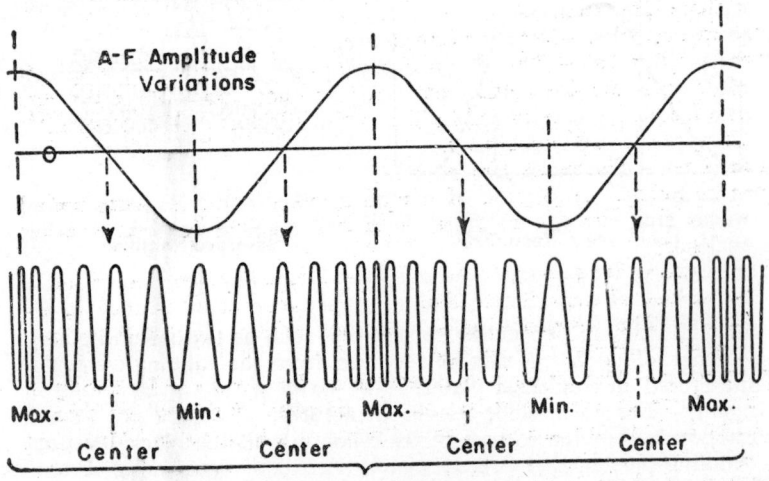

Frequency Modulation

frequency modulation.—A method of radio and television sound transmission and reception in which the carrier is varied in frequency at a rate and to an extent corresponding to audio signals which are to be transmitted and received as modulation. The extent of frequency change in cycles or kilocycles from the average carrier frequency is called deviation and is proportional to changes of amplitude in the audio modulation. The number of frequency changes per second corresponds to the audio or modulating frequency. Average or unmodulated carrier frequency is called the center frequency.

frequency relay.—A relay which operates upon a change of frequency.

frequency response.—The changes which take place in the degree of amplification or attenuation furnished by a device as the frequency of the input power is varied throughout the working range. Usually shown by a graph. Compare *frequency distortion*. See illustration.

frequency run.—A test of characteristics and *attenuation* of a transmission line at various audio frequencies.

frequency tolerance.—The greatest frequency by which a transmitter's *carrier wave* is allowed to vary from the assigned value.

frictional electricity.—*Electrostatic charges* produced by friction or relative movement between two different substances.

frictional machine.—Apparatus for producing *electrostatic charges*

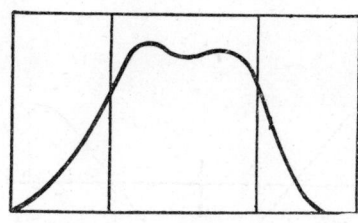

Frequency Response   Full-wave Rectifier

through rubbing two unlike substances, such as leather and glass, together. The charge may be removed from the rubbing elements and stored on condenser plates.

fringing.—Edging in color television pictures.

front porch.—In the composite television signal, the black level or blanking portion that precedes each horizontal sync pulse. Front and back porches together form the pedestal.

full-track recording.—Single-track recording.

full-wave rectifier.—A rectifier that utilizes both the positive and the negative alternations of supply voltage in production of rectified current of single polarity. Conduction is through one rectifier element or through one of two plates in a tube during one alternation and through a second element or plate during the opposite alternation of supply voltage.

fundamental frequency.—1—The frequency of which all other frequencies in a wave are multiples. Compare *harmonic*. 2—In an antenna circuit, the lowest frequency at which the circuit is *resonant* when there is no added capacity or inductance.

fundamental tone.—The *fundamental frequency* in a sound wave.
fundamental wavelength.—The wavelength corresponding to the *fundamental frequency*.
fuse.—A short length of conductor which will carry a certain current continuously but which will melt and open the circuit with an overload applied for a few minutes.
fuse block.—An insulating support provided with terminal connections and used for placing fuses in one or more circuits.

G

G.—Symbol for *electrostatic stress*.

G. or g.—Symbol for conductance in mhos.

gain.—1—The ratio of the output power, voltage or current to the input power, voltage or current in an amplifying system. Generally measured in *decibels*. 2—Gain of *resonant circuit:* The ratio of the voltage developed by the *oscillatory current* flowing through the *capactitive reactance* to the exciting voltage applied to the circuit. 3—Antenna gain.

Galvanometer

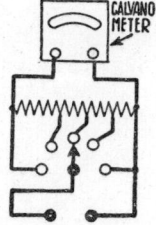

Galvanometer Shunt

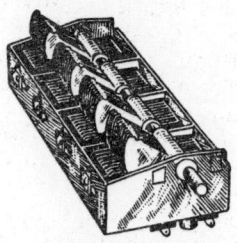

Gang Condenser

galvanometer.—1—An instrument for detecting the presence, the relative intensity, and sometimes the polarity of small currents. 2—Any current measuring instrument not calibrated in amperes or fractions of an ampere, but calibrated in an arbitrary scale.

galvanometer shunt.—A resistance connected across the terminals of a galvanometer and so arranged that a current to be measured may be passed through various definite fractional parts of the total resistance, thus allowing the measurement range of the meter to be extended. *See illustration.*

gamma (γ).—Greek letter symbol for *conductivity*.

gamma rays.—Secondary *radioactive rays* produced by change in rate of motion of *alpha* or *beta rays*. Rays emitted by radioactive substances and having penetrating properties even greater than those of X-rays. Frequencies lying in and above the highest X-ray frequencies.

gang condenser.—Two or more variable *tuning condensers* operated together from one control. *See illustration.*

gang socket.—Two or more tube sockets in one mechanical unit.

gas amplification.—Increase of current in a gas phototube over that in an otherwise similar vacuum phototube, the illumination and anode voltage being the same for both tubes. The increase of total emission due to ionization of gas.

gas phototube.—A phototube whose envelope contains a small quantity of inert gas which ionizes to increase the current and sensitivity of the tube.

gaseous conduction.—Flow of current through an ionized gas.

gaseous conduction rectifier.—A rectifier tube in which a comparatively large current flow results from *ionization* of a gas inside the bulb.

gaseous tube.—A tube having a certain volume of some inert gas such as argon or helium in its bulb, the purpose of the gas being to allow *ionization*. Such tubes are used as *rectifiers* and as *detectors*.

gassy tube.—A tube of the vacuum type in which there is an imperfect vacuum, the remaining gas allowing *ionization*, irregular variations of plate current and a flow of excessive *grid current* even with a negative bias.

gate.—To allow electron conduction through a tube only during definite regularly recurring periods, usually by application to the control grid, screen grid, or plate of voltage pulses whose value and polarity allow conduction, with the gated tube maintained at plate current cutoff while such pulses are not present.

gated automatic gain control.—Keyed automatic gain control.

gated beam tube.—A tube in which electron flow is confined to beams between cathode and plate and in which the elements between cathode and plate include a limiter grid, an accelerator grid, and a quadrature grid. A small change of voltage on the limiter grid causes maximum change of plate current, with further changes of limiter voltage having small effect on plate current.

gauss.—The C. G. S. unit of *magnetic flux density*, equal to one line of magnetic force per square centimeter or equal to 6.45 lines per square inch of cross section.

Geissler tube.—A glass tube containing any of a variety of gases in which a *glow discharge* is accompanied by a color characteristic of the gas.

geometric mean.—The square root of the product of two quantities.

geometric value.—A quantity which depends on the shape and size of parts without reference to their material or to associated parts.

geophysical.—Relating to the earth.

German silver.—A resistance metal consisting of copper, nickel and zinc in various proportions. Nickel silver.

germanium diode.—A crystal of the mineral germanium on which rests the tip of a small diameter conductor. Electrons flow quite freely from germanium, the cathode, to the anode conductor, but with difficulty in the opposite direction. Thus the combination becomes a rectifier for small currents. It is capable of handling very-high frequencies efficiently.

getter.—An alkali metal introduced into the vacuum space of a tube and vaporized for the purpose of absorbing any gases which may have been released in the bulb after sealing off.

ghosts.—In television pictures, appearance on the right of vertical or sloping lines of weaker similar lines. Actually there are two pictures, one produced by normally received carrier signals and a second by signals which have been reflected to the receiving antenna.

gilbert.—The C. G. S. unit of *magnetomotive force*. Equal approximately to 0.796 ampere-turn.

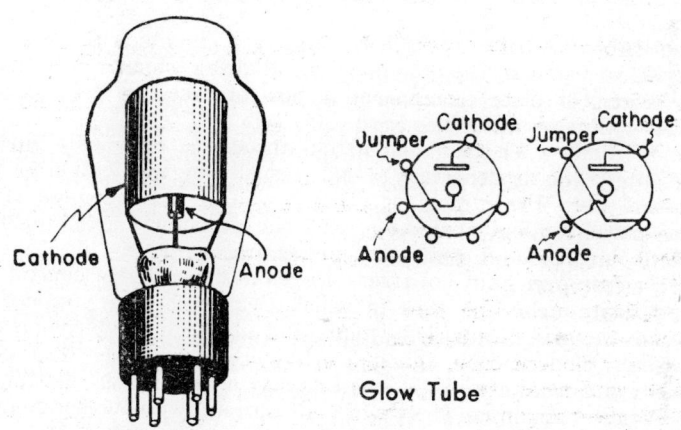

Glow Tube

gimmick.—A small capacitance formed by twisting together two wires, with at least one of them insulated. The insulation acts as dielectric.

glow discharge.—1—A luminous glow which accompanies *ionization* in a gas through which current is passing between electrodes.

glow tube.—Any tube whose envelope contains gas in which there is ionization accompanied by visible glow during electron flow through the tube. One type of voltage regulator is a glow tube.

G_m.—The symbol for *mutual conductance*.

gram.—A unit of weight in the metric or C.G.S. system. Equal to 0.03527 avoirdupois ounce, 15.43 grains.

gram-calorie.—A *calorie*.

graph.—A *characteristic;* a curve showing relations between varying quantities.

graphite.—A soft form of carbon, often used as resistance material.

GREEK LETTER SYMBOLS

Greek letter symbols.—The meanings of Greek letter symbols are given under the English spellings of the names of these letters in the regular alphabetic order. The following letters are those in general use:

gamma (γ) theta (θ) rho (ρ)
delta (δ) lambda (λ) tau (τ)
epsilon (ϵ) mu (μ) phi (ϕ)
eta (η) pi (π) psi (ψ)
      omega (ω)

green gun.—In a three-gun color television picture tube, the electron gun whose beam is intended to excite only the phosphor dots that emit green light.

grid.—A tube electrode having openings through which the electron stream may pass. A *control grid*, a *screen grid*, a *cathode grid*, etc.

grid battery.—A battery which furnishes a *grid bias* voltage. A C-battery.

grid bias.—The direct component of the *grid voltage*. The direct current potential difference between a grid of a tube and *(a)* the tube's cathode, *(b)* the center of the filament in an A.C. filament tube or *(c)* the negative end of the filament in a battery operated filament tube. The symbol for the bias voltage at its source is E_o

grid-bias battery.—A *C-battery*.

grid-bias detector.—A detector tube utilizing the principle of *plate current detection*.

grid capacitance.—The sum of the separate *direct capacitances* between the grid of a tube and all the other elements in the tube. In a three-element tube, the sum of the *grid-filament capacitance* and the *grid-plate capacitance*. The symbol is C_g.

grid-cathode capacitance.—The *electrostatic capacity* between the grid and the cathode of a tube. The symbol is C_{gf}.

grid characteristic.—A graph showing the effect of *grid voltage* changes on *grid current*, the other electrode voltages remaining unchanged.

grid circuit.—That part of a tube circuit included between the cathode and grid, both inside and outside the tube.

grid conductance.—The ratio of a small change in *grid current* to the change of *grid voltage* which produces it, the plate voltage remaining unchanged.

grid current.—What is called positive grid current consists of electron flow from cathode to grid inside a tube when the grid is positive with reference to the cathode, or sometimes when slightly negative. This electron flow is from grid to cathode through external circuits, so the external voltage drop tends to make the grid negative or less positive. Negative grid current is due to electron emission from a grid which has acquired an emissive coating, or may be due to ionization in a gassy tube, or sometimes to internal leakage. Negative grid current electrons flow in external circuits from cathode to grid and voltage drop tends to make the grid positive. Negative grid current may be called reverse grid current.

grid dip meter.—An instrument containing an oscillator, energy from whose plate circuit passes into any externally coupled circuit that is resonant at the oscillator frequency. Increased load on the oscillator reduces its grid current, which flows in and is indicated by a meter. A dip of the meter pointer occurs when oscillator and external circuit are tuned to the same frequency.

grid emission.—*Electron emission* which takes place from a grid electrode which has become overheated.

grid-filament capacitance.—The *direct capacitance* between the grid and the filament of a vacuum tube. In a three-element tube, one-half of the quantity found by adding the *grid capacitance* and *filament capacitance*, and subtracting the *plate capacitance*.

grid-glow tube.—A gas filled tube which contains a cold cathode, an anode, and a grid electrode which electrostatically shields the anode. Ionization allows formation of a *glow discharge* carrying a small current between grid and cathode and, if the negative charge of the grid is reduced to a critical value, the discharge transfers to the anode-cathode path, at the same time becoming sufficiently intense to carry a current many times greater than the original current in the grid circuit. Control of the grid's negative charge through an external leakage impedance, such as the effective resistance through a photocell, allows operation of the grid-glow tube as a relay for the photocell or other external variable impedance. See also *hot cathode grid-glow tube*. *See illustration.*

grid leak.—A resistor connected directly or indirectly between the grid and the cathode of a tube for the purpose of determining or affecting the *grid bias*, and for allowing escape of excess negative charges from the grid. Used with *grid current detection, resistance-capacity coupling*, etc.

grid leak detector.—A detector tube employing the principle of *grid current detection*.

grid modulation.—*Modulation* effected by audio frequency voltages applied to the grid of an oscillator tube in the transmitter circuit.

grid-plate characteristic.—A graph showing the relation between *grid voltage* and *plate current* in a vacuum tube. The mutual characteristic. *See illustration.*

grid-plate transconductance.—*Mutual conductance;* the ratio of plate current changes to grid voltage changes.

grid rectification detector.—A detector tube employing *grid current detection*.

grid resistance.—The effective resistance through the space between the grid and the cathode or filament of a tube. The resistance depends on the *grid bias*. The symbol is R_g.

grid return.—The connection to the cathode or filament of a tube through which it would be possible for grid current to flow. The polarity and potential of the grid return affect the *grid bias*.

GRID SWING

grid swing.—The total variation of *grid voltage* caused by a signal; it is double the peak voltage of the signal.

grid voltage.—The potential difference between a grid and some particular points of the cathode in a tube. The combined effect of the grid bias and the signal voltage. The symbol is E_g or e_g. See *grid bias*.

grid winding.—A transformer winding which forms a part of a *grid circuit*.

grinder.—A form of *static*.

ground.—1—The metallic framework and all parts conductively connected to it in a radio device. 2—A connection to such a framework or a short circuit to it. 3—The lower portion of a *capacity antenna*. 4—The earth and all parts conductively connected to it.

ground absorption.—The power dissipated from a radio wave in the ground or earth.

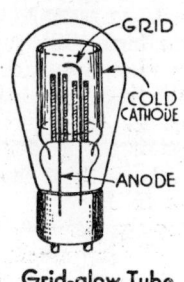

Grid-glow Tube

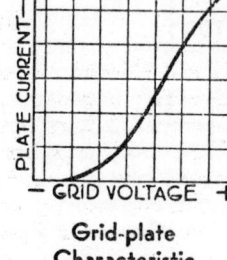

Grid-plate Characteristic

Ground Clamp

ground clamp.—A metal clamp through which connection is made to an earth ground, usually to a water pipe. *See illustration*.

ground potential.—Zero voltage or the earth's voltage. The voltage of the framework or ground connections in a radio device, which may be higher than zero referred to the earth.

ground resistance.—The resistance of the ground or earth portion of an *antenna* system.

ground return circuit.—A circuit having one side carried through insulated metallic conductors and the other through the *ground*.

ground switch.—An *antenna switch*

ground system.—The part of the *antenna circuit* on the ground side of the transmitter circuits and including the earth ground.

ground wave.—The portion of a transmitter's radiated wave which travels along the earth's surface.

ground wire.—1—A conductor connecting with the earth or ground. 2—In a transmitting antenna system, a network of metallic conductors buried under the earth's surface in the space under the aerial.

GROUNDED BASE CIRCUIT

grounded base circuit.—A transistor circuit with which the base element is in both the emitter input and the collector output circuits, much as the grid of a grounded grid amplifier tube is in the input and output circuits.

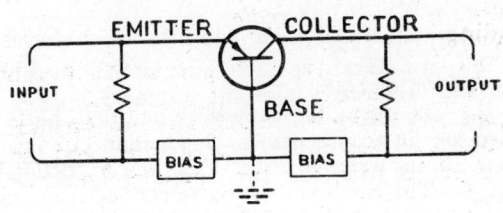

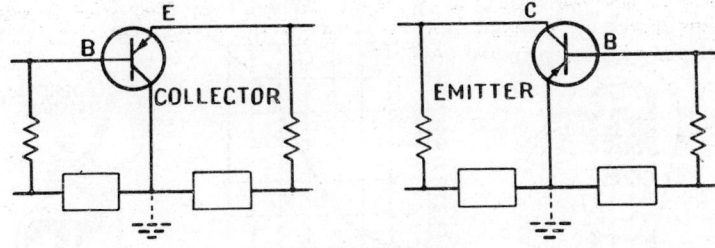

Grounded Base

Grounded Collector   Grounded Emitter

grounded cathode amplifier.—An amplifier tube whose cathode is directly or indirectly grounded, but in any event carries currents or voltages which flow also in both the input and the output circuits. The most widely used amplifier circuit.

grounded collector circuit.—A transistor circuit with which the collector is in the input base circuit and also in the output emitter circuit, much as the plate of a cathode follower tube is in both the input and the output circuits.

grounded emitter circuit.—A transistor circuit with which the emitter is common to the input base circuit and to the output collector circuit. The arrangement is somewhat similar to that of a grounded cathode tube circuit.

grounded grid amplifier.—An amplifier tube whose control grid is grounded for signal currents, either directly or through a capacitor. Input is between cathode and ground, while output is from between plate and ground.

grounded plate amplifier.—A name for a cathode follower.

grounding.—See *ground*.

group frequency.—The number of complete trains of *damped oscillations or waves* occurring in one second.

grown junction.—In a transistor crystal, formation of n-type and p-type regions by introducing appropriate impurity elements during formation of the crystal.

guided wave radio.—*Wired radio*.

Gunn altimeter.—A *capacity altimeter*.

gutta-percha.—A natural vegetable gum similar to rubber and used for insulation. Dielectric constant 3.0 to 5.0.

gyro compass.—A device which indicates geographical direction by action between an electrically rotated wheel and the earth which is rotating on its axis, the two spinning bodies acting as two gyroscopes.

G-Y signal.—A color-difference signal for color television. The green-minus-luminance signal representing primary green coloring minus the luminance or Y-signal. When combined with a plus luminance or Y-signal, outside or inside the picture tube, the G-Y signal yields a primary green signal.

H

H.—A symbol for *magnetizing force*.
h.—Symbol for *henry* (of inductance).
half-track recording.—Dual track recording.
half-wave dipole.—A dipole antenna whose electrical length is equal to a half-wavelength at the carrier frequency for which the antenna is resonant. Electrical length is slightly more than length in inches or feet.
half-wave rectifier.—A *rectifier* which prevents the passage of alternations of one polarity while allowing those of the opposite polarity to pass into its output circuit, thus producing a pulsating direct current. *See illustration*.
hand capacity.—*Body capacity*.

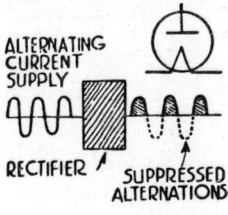

Half-wave Rectifier

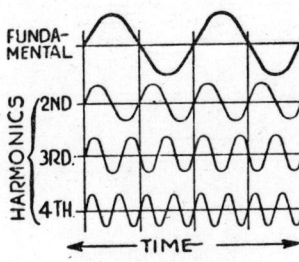

Harmonic

hard drawn copper wire.—A copper wire having high tensile strength.
hard rubber.—Rubber which has been vulcanized at high temperatures and pressures. Used chiefly for insulation. Dielectric constant 2.5 to 3.5.
hard tube.—A vacuum tube which has been highly evacuated and within the bulb of which but little gas remains. The action is entirely thermionic, no intentional *ionization* taking place.
harmonic.—An alternating quantity or a wave motion of which the frequency is an odd or even multiple of a lower frequency called the fundamental. A second harmonic is of double the fundamental frequency, a third harmonic of three times the fundamental frequency and so on. Harmonic and fundamental waves may exist together to form an irregular or *complex wave*. Compare *subharmonic*. *See illustration*.

harmonic amplifier.—An amplifier with the tube so biased as to distort the applied frequency and produce strong *harmonic frequencies* in the output. The desired harmonic is separated by filtering or by passing it through another amplifying stage tuned to its frequency. The process may be carried through several stages, ending with one of the higher harmonics of the fundamental frequency.

harmonic analyzer.—A device for separating *complex waves* into the component *sine waves* of which they are formed.

harmonic current.—A *sine wave* current.

harmonic distortion.—A type of *distortion* in which the output of a tube or circuit contains not only the input frequencies, but also harmonic frequencies of the input. *Wave form distortion*. See *illustration*.

harmonic frequency.—A frequency which is a multiple of another lower frequency called the fundamental; both the fundamental and the harmonic frequencies existing together in a circuit.

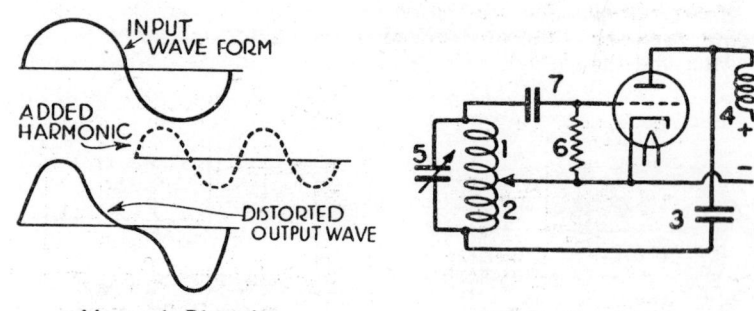

Harmonic Distortion   Hartley Oscillator

harmonic generator.—A circuit which, upon application of sine wave voltage, carries a distorted waveform whose component frequencies include various harmonics of the sine-wave frequency. Used in connection with a harmonic selector for some methods of ultra-high frequency television reception.

harmonic selector.—A circuit tuned to resonance at one of the harmonic frequencies in another circuit which is connected or coupled to the selector circuits, and acts as a signal source. The selector circuit then carries strong oscillatory currents at the harmonic frequency.

harmonic suppressor.—A circuit which reduces the strength of carrier frequency harmonics in the antenna circuit of a transmitter, usually by the use of *band exclusion filters*, for the purpose of limiting the radiation of carrier harmonics.

harmony.—A pleasing combination of tones.

harness.—Wires, cables and terminal connections arranged in such

HARTLEY OSCILLATOR

manner that they may be removed as a unit from the apparatus with which they are regularly used.

Hartley oscillator.—An oscillator whose resonant circuit includes a single capacitor and a tapped inductor in parallel, with part of the inductor between cathode and plate of the tube, and the other part between cathode and grid. There is inductive feedback from plate to grid.

haywire.—Descriptive of apparatus in poor condition or of carelessly made temporary connections.

headphone.—One or two small *telephone receivers* with suitable ear pieces attached to a band or harness going over the head and holding the receivers over the ears.

head set.—A pair of *headphones* attached to a band for support on a listener's head. *See illustration.*

heat loss.—The power which is dissipated from an electrical system in the form of heat. The I^2R *loss.*

heater.—A tube part which is itself heated by flow of current through it and which heats a cathode to a temperature required for *electron emission.*

heater bias.—A difference of potential maintained between the heater and the cathode of a *heater tube* for the purpose of reduc-

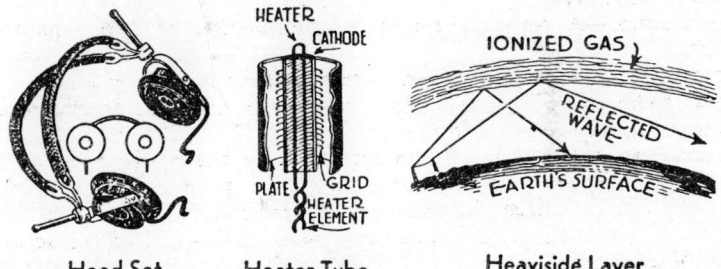

Head Set   Heater Tube   Heaviside Layer

ing the hum tendency. The heater is usually positive with respect to the cathode.

heater circuit.—All of the parts through which flows current passing through a tube heater.

heater current.—The current flowing through the cathode heating element of a tube. The symbol is I_h.

heater resistance.—The *ohmic resistance* of a tube *heater*. The symbol is R_h.

heater tube.—A tube in which the electron emitting *cathode* is electrically separate from the heating element. Within the cathode is a conductor heated to incandescence by alternating current or direct current flowing through it, heat then passing from this element into the cathode. *See illustration.*

heater voltage.—The potential drop across the *heater* of a tube. The symbol for the voltage of the source is E_a, and for voltage across the heater is E_h.

Heaviside layer.—A layer of ionized and somewhat conductive gas existing near the upper part of the earth's atmosphere, the *ionization* being produced by radioactive rays from the sun. This layer reflects and refracts radio waves from a transmitter back toward the earth. Also called the Kennelly-Heaviside layer. *See illustration.*

hecto-.—A prefix meaning one hundred times the unit.

Hefner lamp.—A lamp used as a standard of *luminous intensity;* equal approximately to 0.88 candlepower

helix.—A coil wound in a spiral form.

Helmholtz resonator.—An *acoustic resonator*.

henry.—The practical unit of *inductance*. The inductance in which a current changing its rate of flow one ampere per second induces an electromotive force of one volt. Abbreviated *h*.

Hertz antenna.—An antenna circuit in which the ground forms no essential part. A *capacity antenna* employing a counterpoise instead of a ground. *See illustration.*

Hertzian waves.—*Radio waves*. Waves of frequencies from the lowest used in radio transmission up to the lower limits of infrared or heat rays.

heterodyne.—A separate force; a combination of two separate forces.

heterodyne frequency meter.—An *autodyne frequency meter*.

heterodyne interference.—*Beat interference*. Also, interference caused by *re-radiation* from a receiver.

heterodyne reception.—Radio reception by the combination of a received frequency with another frequency produced in an oscillatory circuit within the receiver. *Superheterodyne reception* and *autodyne reception*. A form of reception making use of the action of *beating*.

Heusler's alloy.—An alloy metal containing copper, aluminum and manganese. It forms a magnetic material.

h-f.—Abbreviation for *high-frequency*.

high band.—The very-high frequency television broadcast channels numbers 7 to 13 inclusive.

high-fidelity.—Descriptive of a sound reproducing system capable of audible output which faithfully follows input signals. Among desirable properties of such a system are: Minimum practicable distortion of any kind. No audible hum. Negligible noise of any kind. Small phase shifts between input and output. Nearly constant power gain and no overloading on strongest signals. Frequency range from 40 or 50 cycles up to about 15,000 cycles per second.

high frequency.—By definition, any frequency from 3 to 30 megacycles per second.

HIGH-FREQUENCY COMPENSATION

high-frequency compensation.—Means for extending the gain of a broad band amplifier to some desired high limit, usually by reduction of the shunting effect of circuit and tube capacitances and by use of peaking inductors to increase amplifier load reactance and gain as frequency increases.

high frequency resistance.—The effective resistance in a high frequency circuit; the total of all effects which result in energy losses. High frequency resistance includes losses due to *eddy currents, skin effect, distributed capacity, dielectric losses* and *ohmic resistance.*

high frequency transformer.—A *radio frequency transformer.*

high-mu tube.—A tube having a high *amplification coefficient.*

high pass filter.—A filter which allows comparatively free passage through it of frequencies above a certain *cutoff frequency*, but which greatly attenuates all lower frequencies. *See illustration.*

high resistance voltmeter.—A voltmeter having resistance suffi-

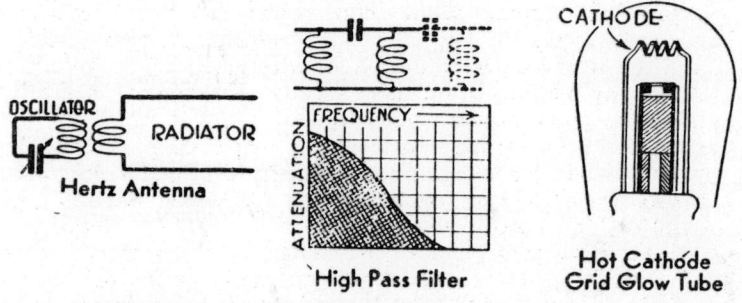

Hertz Antenna   High Pass Filter   Hot Cathode Grid Glow Tube

ciently great that very little current is drawn by the meter from the circuit in which potential is measured. Generally a *moving coil instrument* taking not more than one milliampere of current for its own movement.

hill and dale recording.—Vertical recording.

hold control.—A television receiver adjustment which alters free running frequency of the horizontal or vertical sweep oscillator by changing the time constant of a capacitor-resistor circuit. Such adjustment allows the oscillator to be synchronized by received signals, for proper timing of electron beam deflections in the picture tube.

hole.—An atom in a p-type transistor crystal which temporarily lacks its normal complement of negative electrons, and thus is positive. When this atom regains an electron which has come from another atom, the hole (positive charge) will have moved to that other atom. It is in this manner that holes or positive charges move through the crystal.

HOMODYNE RECEPTION

homodyne reception.—*Zero beat reception.*

hook transistor.—A four-element transistor in which are three junctions between n-type and p-type crystal regions and in which current gain may be relatively great due to alpha being greater than unity or greater than 1.00.

hook-up.—The circuit of an electrical device, or a diagram showing the circuit.

hook-up wire.—Tinned copper wire, solid or stranded, with either fabric or rubber insulation which is removed easily.

horizontal blanking.—Between successive horizontal lines in television pictures, the time period during which composite signal voltage remains at the black level except for a horizontal sync pulse, and during which the electron beam should be cut off by black level voltage. Horizontal retrace occurs within the horizontal blanking period.

horizontal retrace.—Descriptive of voltages and currents which, were the electron beam of a television picture tube not blanked, would cause the beam to move from right to left between successive picture or raster lines. The sudden reversal in sawtooth voltages and currents for deflection corresponds to the retrace.

horizontal sweep.—Deflection of the electron beam from side to side in a television picture tube or cathode-ray tube. Descriptive also of any portions of a receiver or instrument in which are voltages or currents causing such deflection.

horn.—A tubular member of varying shape attached to a loud speaker driver for assisting the radiation of sound waves.

horsepower.—A unit of *power* equal to 746 watts. A rate of work equal to 33,000 foot-pounds per minute.

horse shoe magnet.—An electromagnet or a permanent magnet in which the poles or ends of the *core* are brought close together.

hot cathode grid-glow tube.—A *grid-glow tube* having as a primary source of electrons a heated cathode, the resulting ability to handle large currents allowing this tube to carry considerable power directly between its elements. These tubes are filled either with mercury vapor or with neon gas. Compare *grid-glow tube*. *See illustration.*

hot cathode rectifier.—Generally a *mercury vapor rectifier.* Any *thermionic* or *gaseous conduction rectifier* obtaining primary electron emission from a heated cathode.

hot cathode X-Ray tube.—A tube from which *X-rays* are secured by electron emission from a heated filament.

hot chassis.—In any receiver or instrument, a metal support or chassis conductively connected to one side of the building power line, therefore at line voltage with respect to anything connected to the other side of the line.

hot side.—The high voltage portion or side of an electric circuit.

hot wire meter.—1—A current measuring instrument in which movement of the pointer is allowed by expansion and lengthening

of a conductor carrying the current, the expansion being due to
heating of the conductor by the current being measured. *See
illustration.* 2—A *thermocouple instrument* in which the thermocouple is heated by a separate conductor carrying the current to
be measured.

howling.—A sound which is the result of audio frequency *feedbacks*
or *acoustic feedbacks* causing vibration of the elements in an
amplifying tube.

H-section attenuator.—An *attenuation network* having its elements
arranged in the manner of the capital letter "H"; a *constant impedance unit*. *See illustration.*

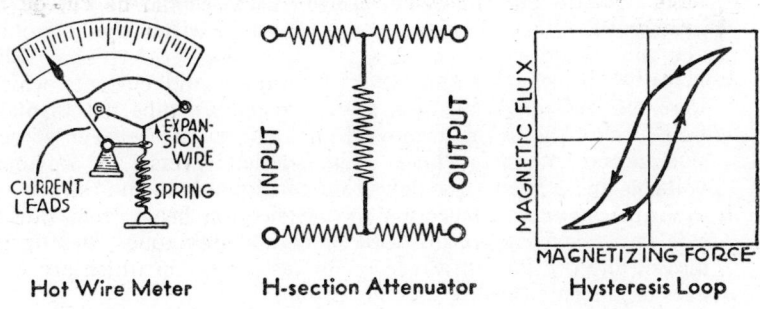

Hot Wire Meter   H-section Attenuator   Hysteresis Loop

hue.—Descriptive of a color sensation which results from certain
wavelengths of light energy and which may be described as blue,
green, red, and so on. Hue is not altered when a given color is
lighter or darker nor by changing the intensity of colored light,
but depends entirely on the predominant wavelength of radiant
energy.

hue control.—In a color television receiver, an adjustment that
alters the phase of color oscillator output voltages with respect to
phase of the burst, and which thus alters the hues or wavelengths
of light in reproduced pictures.

hum bars.—One or two pairs of alternate dark and light horizontal
bands across television pictures. The bands result from combination with video signals of voltages which would cause hum in a
sound receiver, usually at power line frequency or ripple frequency.

hum bucking.—Reduction of audible hum in alternating current
receivers by combining with the hum-producing voltage another
voltage of the same frequency but opposite in phase.

hyp.—A unit of *attenuation*.

hysteresis.—1—A delay or lag in the increase of *magnetic flux* with
increase of the *magnetizing force* and the similar lag in decrease
of flux with a decrease of the force; the rates of increase and of
decrease in flux not being the same with a rise of magnetizing force

as with a falling off in force. 2—*Dielectric hysteresis*.

hysteresis coefficient.—The energy in ergs dissipated in one cubic centimeter of iron by the *hysteresis* effects during one cycle of *magnetizing force*.

hysteresis loop.—Two curves meeting at their ends and forming a loop; one of them showing the rise and the other showing the fall of *magnetic flux* during one cycle of an alternating *magnetizing force*. The power loss due to *hysteresis* is roughly proportional to the area of the loop. *See illustration*.

hysteretic.—Pertaining to the effects of *hysteresis*.

hysteretic angle.—The phase difference from *quadrature* caused by hysteresis.

hysteretic loss.—The amount of power in watts dissipated because of *hysteresis*.

I

I.—Symbol for *effective* value of current in amperes.
i.—Symbol for *instantaneous* value of current in amperes.
Iconoscope.—A television camera tube in which an area called the mosaic carries minute photosensitive elements, a conductive plate, and a collector. Lights and shadows of an image focused on the mosaic result in varying voltage.
icw.—Abbreviation for *interrupted continuous waves*.
I-demodulator.—A color television demodulator in which combine the chrominance signal and a voltage from the color oscillator to produce an I-signal.
idle current.—The *reactive current*.
I. E. C.—Abbreviation for *International Electrotechnical Commission*.
i.f. or i-f.—Abbreviation for *intermediate frequency*.
I_f.—Symbol for *filament current*.
I_g or i_g.—Symbol for *grid current*. For current through the grid nearest the cathode in a multigrid tube the symbol is I_{g1}, and for the grid second from the cathode it is I_{g2}.
ignitron.—A gas-filled tube (or tank) with mercury pool cathode. A controlled rectifier in which starting of electron flow is controlled by an ignitor element.
I_h.—Symbol for *heater current*.
I_j.—Symbol for current in a tube electrode.
illumination.—The volume of light reaching a body. The density of *luminous flux*. Units are the *foot-candle,* the *lux* and the *phot*.
I_m.—Symbol for *maximum* value of current.
image.—An optical reproduction or picture of objects formed by *reflection, refraction* or transmission of light. A *real image* or a *virtual image*.
image dissector.—A television camera tube which converts lights and shadows of a focused image into corresponding changes of voltage.
image effect.—An increase in the *effective height of an antenna* due to its ground system.
image frequency.—The carrier frequency of an undesired signal which is capable of combining with the oscillator frequency in *superheterodyne reception* to produce the intermediate frequency. The image frequency is as far above the oscillator frequency as the desired carrier frequency is below that of the oscillator, differing from the desired carrier by twice the intermediate frequency.

IMAGE IMPEDANCE

image impedance.—The impedance which, when used at the end of a transmission network allows no *reflection loss* at the junction between network and impedance.

image transfer constant.—In a transmission network terminated in its *image impedances,* one-half the natural logarithm of the vector ratio of the apparent power entering the network to the apparent power leaving the network. The factor by which power is reduced in a network.

imaginary component.—The *reactive current or component.*

impact excitation.—*Impulse excitation.*

impedance.—Opposition to flow of alternating current, measured in ohms. The symbol is Z or z. Impedance is the combined effect of resistance and reactance. In the same circuit with resistance may be only inductive reactance, only capacitive reactance, or both kinds. Impedance in ohms is equal to the quotient of dividing applied alternating volts by amperes of resulting alternating current. Because of phase shifts in inductances and capacitances the impedance is not equal to the sum of opposition which would be due to resistance and reactances considered by themselves, but is equal to the square root of the sum of the squares of resistance and either kind of reactance when only one kind is present. If both inductive and capacitive reactance are present, impedance is equal to the square root of the sum of the square of resistance and of the square of the difference between the two reactances. If the two reactances are equal their difference is zero, they cancel, and impedance becomes equal to resistance alone, this being the condition of resonance.

impedance coupling.—Coupling obtained by making a single impedance a part of each of the coupled circuits. *Direct coupling.*

impedance factor.—The ratio of the *impedance* to the *ohmic resistance* in an alternating current circuit.

impedance match.—The condition wherein the impedance of a connected load is equal to the internal impedance of the source, or the condition wherein a coupling device between source and load has a ratio of its input to its output impedances equal to the ratio of impedances of the source and the load. The condition with which there is maximum power transfer from source to load. *See illustration, page following.*

impedance triangle.—A right angle triangle in which, when the lengths of the two sides are proportional to the *resistance* and the *reactance,* the length of the hypotenuse is proportional to the impedance of a circuit wherein inductive and capacitive reactances are in series.

impedance wave trap.—A *parallel resonance* circuit in series with an antenna circuit; the greatest impedance being offered to currents of an undesired frequency to which the trap circuit is tuned. *See illustration.*

impregnation.—Filling the spaces between conductors, laminations

IMPRESSED VOLTAGE OR E.M.F.

or other parts with an insulating or binding material, usually under the action of heat or vacuum.

impressed voltage or e.m.f.—The voltage applied to a circuit or device. The potential difference applied by an external source of e.m.f. to the ends of a circuit or to the terminals of an electrical device.

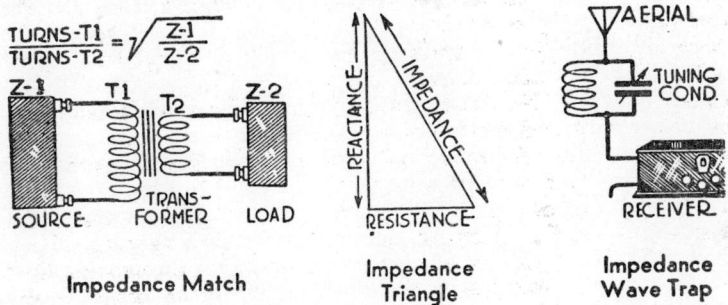

Impedance Match Impedance Triangle Impedance Wave Trap

impulse.—A sudden and brief change in current or voltage in a circuit.

impulse electromotive force.—A voltage which decreases rapidly at the beginning of a current which it causes. A diminishing voltage, the average value of which is taken over a period equal to the circuit's *time constant*.

impulse excitation.—Momentary application of an exciting voltage to an oscillatory circuit in which are produced damped oscillations at the circuit's *natural frequency*, these oscillations continuing after the applied voltage has ceased. Shock excitation.

impurity.—An element introduced into a transistor crystal to make it either n-type or p-type. Atoms of n-type impurities have more valence electrons than crystal atoms, and thereby furnish electrons which may move in the crystal. Atoms of p-type impurities have fewer valence electrons and cause positive charges or holes which may move in the crystal.

incandescence.—Lighting or glowing of a substance because of its heat.

incremental tuning.—A method of television channel selection with which circuits for r-f amplifier, r-f oscillator, and mixer are made resonant at lower and lower frequencies by switching additional sections of tapped inductors into the tuned circuits. Each inductor is electrically continuous, with only the required portion selected for each channel.

index of refraction.—An indication of the amount of bending or *refraction* undergone by light rays upon entering and leaving a medium; equal to the ratio of the rate of travel of light in a vacuum (or in air) to its rate of travel in the medium considered.

INDICATION RANGE

indication range.—The range of an instrument within which values are indicated without reference to accuracy. Compare *measurement range*.

indirect current.—*Alternating current*.

indirect wave.—The *sky wave*.

indirectly heated cathode.—A tube cathode which is heated from a separate conductor carrying the heating current, placed within or close to the cathode but electrically insulated from it. The cathode of a *heater tube*.

indoor antenna.—An antenna system having its aerial wire within a building. *See illustration*.

induced charge.—An *electrostatic charge* produced on a conductor by the field around another nearby conductor.

induced current.—A current produced by changes of *magnetic flux* and the resulting induced voltage.

induced voltage or e.m.f.—A voltage produced by *induction*.

inductance.—The property of a circuit by which a varying current in it produces a varying magnetic field of which the moving lines of force cause voltages in the same circuit or in other nearby circuits. The ratio of the magnetic flux to the current which produces it. *Self-inductance* or *mutual inductance*. Measured in *henrys*.

inductance bridge.—A *Wheatstone bridge* especially arranged for measurement of *inductance* by the balancing out of resistance and capacity in a circuit. *See illustration*.

inductance coil.—A coil possessing *inductance* or the ability to produce voltage in its own winding or in other nearby conductors. The coil may be used to provide *inductive reactance* in a circuit. *See illustration*.

Indoor Antenna

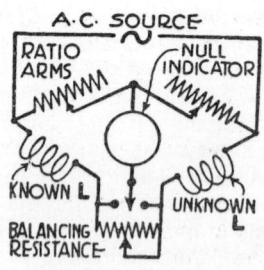

Inductance Bridge

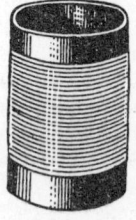

Inductance Coil

inductance switch.—A switch for connecting or disconnecting portions of a winding. *See illustration*.

induction.—The action by which an electrostatic charge, an electromotive force or a magnetic condition is produced in a material by

INDUCTION COIL

magnetic or electrostatic lines of force. *Electrostatic induction, electromagnetic induction* and *magnetic induction*.

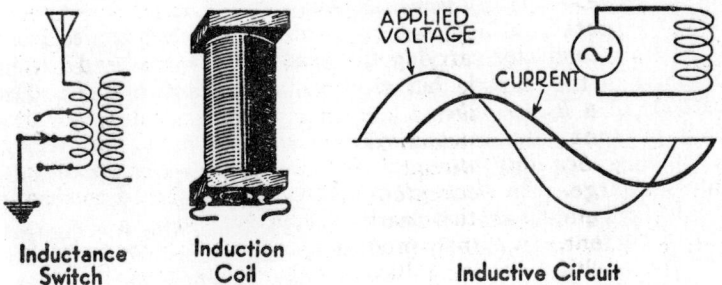

Inductance Switch **Induction Coil** **Inductive Circuit**

induction coil.—1—A coiled conductor, with or without an iron core, which utilizes *electromagnetic induction* to cause changes in a current. *See illustration.* 2—An iron-core coil with primary only or with both primary and secondary windings used to produce sudden changes of current or voltage.

induction density.—*Flux density.*

induction field.—In the electric or magnetic field existing around a conductor, the portion of that field from which the energy is returned to the conductor rather than being radiated as part of the *radiation field.*

induction instrument.—An instrument in which movement of the pointer results from the action of fixed coils on moving parts in which currents are produced by *electromagnetic induction.*

induction loud speaker.—A *moving coil loud speaker* in which the current reacting with the stationary field is induced in the moving member.

induction machine.—An alternating current machine having one stationary and one rotating winding; an *induction generator* or *induction motor*.

induction motor.—A motor using alternating current to produce a magnetic field which revolves within the machine. This moving field induces currents in the windings of a rotor. Reaction between the rotating field and the conductors carrying the induced currents causes the rotor to revolve with the moving field.

inductive capacity.—Specific inductive capacity or *dielectric constant.*

inductive circuit.—A circuit containing considerably more *inductive reactance* than *capacitive reactance,* more inductance than capacity. *See illustration.*

inductive coupling.—A form of coupling in which energy transfer results from *electromagnetic induction;* a single magnetic field

INDUCTIVE DISTURBANCE

passing through the two coupled circuits. Voltage is induced in the one circuit by the moving lines of force which arise from the changing current in the other circuit. Coupling in which a single *mutual inductance* forms a part of the total inductance associated with each of the two coupled circuits. *See illustration.*

inductive disturbance.—Interference produced by voltages picked up from *inductive fields* close to an antenna.

inductive feedback.—A feedback of energy from an output circuit to an input circuit through an *inductive coupling.*

inductive load.—A load circuit in which the *inductive reactance* exceeds the *capacitive reactance.* A load in which the current lags behind the applied voltage. Compare *capacitive load.*

inductive reactance.—The part of the *reactance* which is due to inductance in an alternating current circuit. Measured in *ohms.* Equal to the number of ohms resistance which would have an effect in opposing current flow equal to the effect of the *counter-*

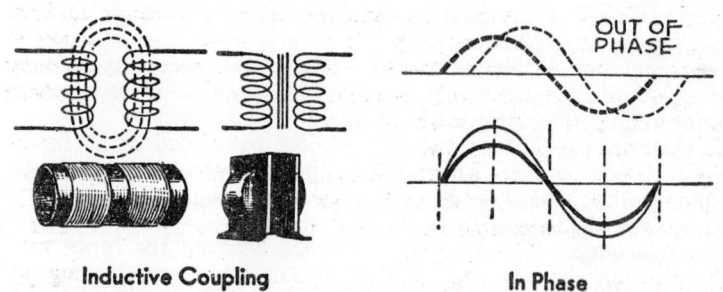

Inductive Coupling   **In Phase**

electromotive force developed by the moving field around the conductor. The symbol is X_l.

inductive susceptance.—The reciprocal of the *inductive reactance* in a circuit containing negligible resistance. Measured in *mhos.*

inductive time constant.—Time required for current in an inductive circuit to reach 63.2 per cent of the value it would reach after an indefinitely long time, the time being measured from the instant at which voltage is applied.

inductor.—In the usual meaning of the word, any coil or winding, with or without an iron core, whose principal purpose is to provide inductance or inductive reactance. Otherwise the inductor may be a wire or other conductor not coiled, used to provide inductance or inductive reactance at high frequencies.

inertia.—The property of continuing at rest when in that condition, or of continuing in motion without change of direction when so in motion. Self-inductance is called electrical inertia.

infinite baffle.—A speaker enclosure having no opening through which may emerge sound waves from the back of a cone or other vibrating member.

infinite impedance circuit.—A name sometimes applied incorrectly to a *parallel resonance circuit*.

infra-red rays.—*Radiant energy* of frequencies lower than those which are visible; rays lower in frequency than the red rays of the visible spectrum and extending down to the highest radio frequencies.

in phase.—Descriptive of two currents, two voltages, or a current and a voltage which are alike in frequency and which pass through their zero and maximum values at the same instant. *See illustration*.

in-phase component.—The *active component*.

input.—1—The terminals of an electrical instrument at which it receives current or voltage from some other instrument. The entering point for incoming current or voltage from some other part. 2—The power, voltage or current absorbed or used in an electrical device.

input admittance.—The reciprocal of the *input impedance*.

input circuit.—The circuit through which power, voltage or current is applied to any electrical device.

input impedance.—The ratio of the alternating voltage applied at the input of a device to the alternating current produced in the input circuit, other impressed voltages being absent.

input reactance.—The *reactance* presented by a load or an input circuit to the source. The ratio of the sine wave input voltage to the resulting sine wave current of the same frequency and ninety degrees out of phase with the voltage.

input resistance.—The *internal resistance* between the input terminals of any device. The ratio of the sine wave input voltage to the resulting sine wave input current which is in phase with the voltage.

input transformer.—A transformer through which voltages are applied to an amplifying stage, a complete amplifier or other part.

inside antenna.—An *indoor antenna*.

instantaneous value.—The value of an alternating power, current or voltage at any one point in its cycle. A value intermediate between zero and the peak value.

Institute rating.—A rating according to the standards of the American Institute of Electrical Engineers.

instrument.—A device for measuring the existing value of a quantity or effect.

insulation.—Material which opposes passage of electric current through it and which is used to electrically separate conductors from other conductors. A non-conductor.

insulation leakage.—The combined effect of *surface leakage* and *volume leakage* of an insulator.

insulation resistance.—1—The *ohmic resistance* of a material to voltage tending to break through it. 2—The net resistance resulting from *volume resistance* and *surface resistance* in parallel.

insulator.—A device or part which separates conductors and sometimes supports them while preventing flow of continuous electric current through itself.

integrating filter.—Between the sync section and vertical sweep oscillator of a television receiver, shunted capacitors and series resistors in which serrated vertical sync pulses build up a capacitor charge and voltage which starts each operating cycle of the oscillator. The filter gathers or integrates charges imparted by successive parts of the serrated vertical pulse.

integrating instrument.—An instrument which records the total quantity of current or the total power used during a period of time.

intelligibility.—The percentage of the total number of ideas sent over a transmission system which may be identified or understood at the far end.

intensifier.—A separate radio frequency *stage of amplification* connected between an antenna and the input to a receiver.

Integrating Filter

intensity.—The quantity or amount of a force or pressure per unit of cross section, area, volume, etc.

intensity control.—A saturation control in a color television receiver.

intensity modulation.—Varying the strength or intensity of the electron beam in a cathode-ray tube by means of voltage applied to its control grid-cathode circuit.

intercarrier beat.—A voltage at 4.5-megacycle frequency due to mixing or beating of video and sound intermediate frequencies in the video detector of a television receiver. If this frequency reaches the picture tube it causes an interference pattern in pictures.

intercarrier buzz.—A sharp buzzing sound from a television receiver employing intercarrier sound. Due to misadjustment of receiver controls or sometimes to overmodulation in received carrier signals.

intercarrier sound.—In television receivers, a 4.5-megacycle center frequency or average frequency which is frequency-modulated with audio signals. This modulated signal appears at the output of the video detector or a separate sound detector wherein video and sound intermediate frequencies beat together to produce their difference frequency of 4.5 megacycles, with frequency modulation of

INTERELECTRODE CAPACITY

sound carriers retained. The 4.5-megacycle frequency-modulated signal is amplified at this frequency, then demodulated to recover the sound portion of programs.

interelectrode capacity.—The *electrostatic capacity* between the elements of a vacuum tube. *Grid-plate capacitance, plate-cathode capacitance, grid-cathode capacitance*, etc.

interference.—1—Radio interference is the effect of any electrical waves or fields, other than a desired signal, in producing confused sounds or other signal indications from a receiver. 2—Sound interference is the effect of two sound waves on each other whereby the sound intensity is raised or lowered. 3—Light interference is the effect on each other of two light beams whereby the resulting beam is weakened or broken into bands.

interference eliminator.—A *wave trap*, a bypassing resistance or other device for reducing the effects of radio *interference*.

interference guard bands.—The additional range of frequencies above and below those occupied by the modulated carrier and the

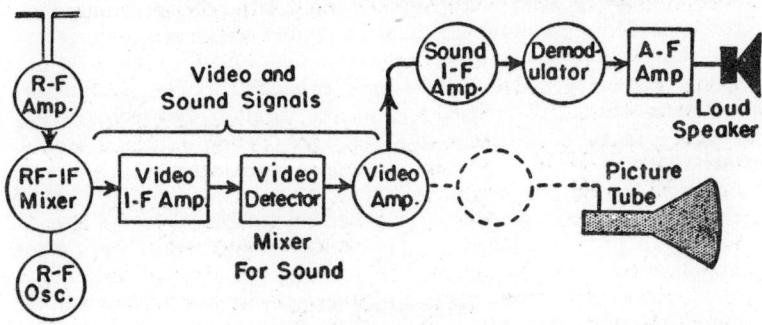

Intercarrier Sound

tolerance frequency, being allowed for avoidance of *interference*.

interference level.—The *static level*.

interlaced scanning.—Reproduction of television pictures by causing the electron beam in the picture tube to trace only alternate horizontal lines during one downward travel and to trace intervening lines during the following downward travel. Each downward travel forms one field, with two successive fields forming one frame or one complete picture. Lines for each alternate field begin at the upper left, those for intervening fields begin midway across the top of pictures.

interleaving.—A method of transmitting and receiving chrominance and luminance signals for color television within the same range of video frequencies. Chrominance signal sidebands concentrate at harmonics of the subcarrier frequency while luminance signal sidebands concentrate at harmonics of the horizontal line frequency. Successive harmonics of opposite kinds are separated by about 8,800 cycles throughout the video frequency range.

intermediate frequency.—1—A frequency higher than that used as *modulation frequency*, but lower than a *carrier frequency*. 2—In *superheterodyne reception*, the frequency produced by beating together of the received modulated carrier and the locally generated oscillator frequency. Abbreviated *i.f.* or *i-f*.

intermediate-frequency amplifier.—An amplifying system of one or more stages in which are strengthened the signals at intermediate frequency applied to this amplifier from a converter or mixer. Amplified signals pass to a detector or demodulator. A television receiver intermediate-frequency amplifier is between tuner and video detector.

intermodulation.—A type of sound distortion occurring in an amplifier or reproducing system whose gain is not at least approximately uniform for all frequencies of the input signal. One effect is beating or mixing of different frequencies of a complex input voltage wave, to produce sum and difference frequencies which are not musically harmonic and which cause harsh or rough output sounds. Also, unnatural and disagreeable effects due to a strong low-frequency input voltage modulating higher-frequency voltages in the same sound program.

internal capacity.—The *electrostatic capacity* between conductors inside an electrical device; such as the capacity between the elements of a tube. *Distributed capacity*.

internal drop.—The potential difference existing inside a source between the output terminals; equal to the e.m.f. acting to send current through a generator or other source.

internal output admittance.—The reciprocal of the *internal output impedance*.

internal output impedance.—The ratio of the alternating voltage across the output terminals of a device to the resulting alternating current between these terminals, there being no other impressed voltages.

internal resistance.—The resistance of that part of a circuit which is inside of a source between its output terminals.

international ampere.—The unvarying current which causes a deposit by electrolysis of 0.001118 gram of silver per second with a silver anode from a fifteen per cent silver nitrate solution as electrolyte.

international candle.—A unit of *luminous intensity*. The light emitted by the flame of a sperm candle seven-eighths inch in diameter burning at the rate of 120 grains or 7.776 grams per hour.

international Morse code.—The *continental code*.

international ohm.—The resistance offered to an unvarying electric current by a column of pure mercury of uniform cross section, 106.3 centimeters long, weighing 14.4521 grams, at a temperature of zero centigrade .

international volt.—The electrical pressure or electromotive force which sends a current of one international ampere through a resistance of one international ohm.

interrupted continuous waves.—A *carrier wave* which is broken up into a series of wave trains, the groups of waves occurring at an audio frequency. The interrupted wave is further broken up into the dots and dashes of the telegraphic *code*. Abbreviated *icw*. *See illustration*.

interstage coupling.—The coupling used between the several tubes in *cascade amplification*.

interstage shielding.—Metal *shields* designed to prevent feedbacks between stages in an amplifying system. *See illustration*.

interstage transformer.—A transformer having its primary connected in the plate circuit of one amplifying tube and its secondary connected to the grid circuit of the tube in a following stage in *cascade amplification*.

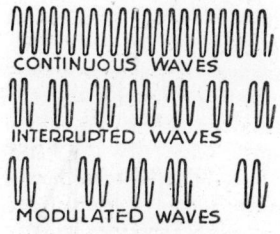

Interrupted Continuous Waves

Interstage Shielding

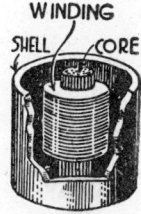

Iron-clad Electromagnet

intrinsic barrier transistor.—A four-element transistor in which atoms in the fourth region or section of the crystal have neither excess nor deficiency of valence electrons. Allows high-frequency operation.

intrinsic brightness.—The total *luminous intensity* of a source in any direction divided by the apparent area of the source viewed from that direction. See *brightness*.

inverse duplex amplification.—A *reflex circuit* in which each reflexed tube carries either a light radio frequency and a heavy audio frequency load, or else a heavy radio frequency and a light audio frequency load.

inverse peak voltage.—*Peak inverse voltage*.

inverse resonance.—*Parallel resonance*.

inverter.—A tube whose principal purpose is to reverse the polarity or produce an opposite phase from an applied waveform. There is inversion between output and input of a grounded cathode amplifier, but not of a grounded grid amplifier or of a cathode follower.

ion.—An extremely minute particle of an element or compound, a gas or a solid, which has acquired a positive charge or a negative charge for which the ion acts as a carrier through a gaseous space or an electrolyte between electrodes. A positively charged ion is called a *cation* and one negatively charged is called an *anion*.

ion burn.—A permanently darkened area near the center of the viewing screen of a television picture tube operated with magnetic deflection. The cause is bombardment of the screen phosphors by ions not separated from the electron beam by an ion trap or otherwise.

ion trap.—Means for preventing concentrated streams of ions from striking the viewing screen in a television picture tube. Electric fields produced in the electron gun bend or deflect both the ions and the electrons. Magnetic fields produced by one or more permanent magnets mounted outside the tube neck deflect electrons but have little effect on ions. Thus ions are separated from the electron stream and trapped within the electron gun.

ionization.—An action by which a gas is made conductive. A result of collisions of *electrons* with atoms of the gas, wherein other electrons are detached from the atoms which then are *positive ions*. This increase of negative electrons and positive ions corresponds to an increase of current passing through the gas. Ionization may be induced by heat, light, electric potential or other conditions which cause greater acceleration of electrons.

ionization potential.—The potential (in volts) required to separate an electron from an atom in the process of *ionization*. The voltage through which an electron must pass to acquire sufficient energy to liberate an electron from an atom.

ionized layer.—The *Heaviside layer*.

I_p or i_p.—Symbol for *plate current*.

IR drop.—The drop in voltage due to *resistance* in a circuit through which current is flowing. Equal to the current times the resistance.

I^2R loss.—The amount of power which is dissipated as heat in a circuit's resistance. Equal to the square of the current times the resistance. Measured in *watts*.

I.R.E.—Abbreviation for Institute of Radio Engineers.

iron-clad electromagnet.—An *electromagnet* placed within a cup of soft iron so that one end of the magnet core and the rim of the cup form the two poles. *See illustration, page preceding*.

iron-core coil.—An inductance coil having a center of iron which increases the *permeability* of the magnetic path and increases the *inductance* of the winding.

iron loss.—The energy dissipated in the iron core of a transformer or choke coil through production of *eddy currents* and by *hysteresis*. The core loss. Measured in *watts*.

I-SIGNAL

I-signal.—One of the parts of a chrominance signal for color television. Sidebands of the plus I-signal center around a color phase which is 57 electrical degrees from the reference phase or burst phase.

isochronous.—Maintaining an unvarying frequency or period.

isolantite.—An insulating and supporting material having small power loss at high frequencies. Dielectric constant 3.6.

isolated.—Descriptive of apparatus not easily accessible to unauthorized persons.

isolation transformer.—A transformer having an insulated secondary winding usually of one-to-one voltage ratio, connected between a power line and a receiver which has a hot chassis to prevent line voltage from reaching the chassis directly. Used during service operations.

iterative impedance.—In a transmission line, the ratio of the applied *electromotive force* to the resulting *steady state* current in a uniform circuit of infinite length or in a circuit of periodically repeating structures. The geometric mean of the open circuit impedance and the short circuit impedance of a recurrent symmetrical structure.

J

jack.—A set of contacts which open or close to change circuit connections upon insertion or removal of a plug carrying the ends of another circuit. *See illustration.*

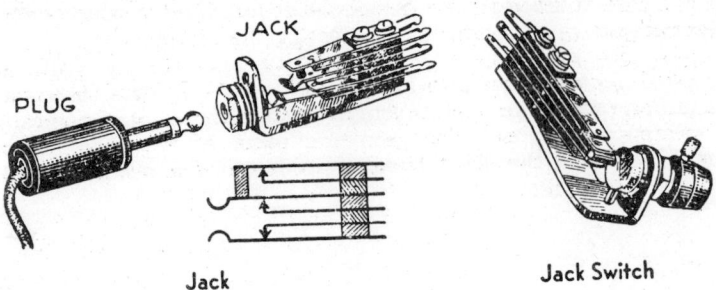

Jack   Jack Switch

jack switch.—A switch of the single or double throw and single- or multi-pole variety with contacts which are opened and closed by means of a cam. *See illustration.*

jamming.—*Interference* from a transmitter which it is not desired to receive.

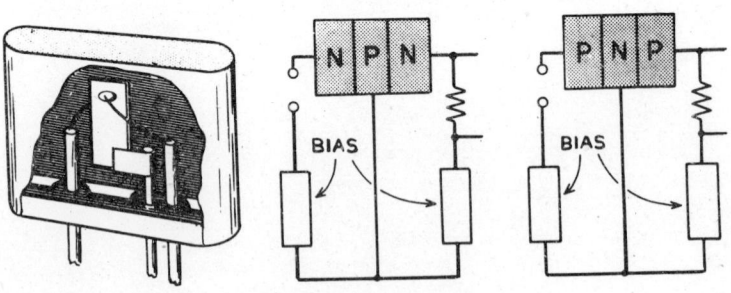

Junction Transistor

JAN.—Abbreviation for Joint Army-Navy.

jar.—A unit of *electrostatic capacity* equal to 0.0011 microfarad or to 1,000 centimeters of capacity.

joule.—The practical unit of electrical *energy* or *work*. The work done by a power of one watt in one second. The work expended in sending one ampere of current through one ohm of resistance for one second. Equal to 10,000,000 *ergs*.

joulean heat.—The heat produced by the current flowing against the *resistance* of a conductor.

Joule's law.—A law which states that the energy in heat units liberated in a circuit or part of a circuit is equal to the product of the current squared, the resistance and the time in seconds.

junction transistor.—A transistor in which n-type and p-type regions are parts of the same crystal structure, which includes emitter, collector, and base, also other regions in some constructions. Distinguished from a point-contact transistor in which some elements are small conductors contacting the crystal.

K

K.—Symbol for *dielectric constant* or specific inductive capacity.
K.—Symbol for resistance of 1,000 ohms.
k.—Symbol for a constant, usually for *coupling coefficient*. Abbreviation for the prefix *kilo-*.
kallirotron.—A type of vacuum tube having the property of *negative resistance*.
Karolus light valve.—Apparatus consisting of a *Kerr cell* between two *Nicol prisms*, used to vary in accordance with a signal the amount of light passing through the valve. The first prism polarizes the light, and the second is in such angular position that it freely passes the polarized beam. Signal voltages applied to the Kerr cell rotate the beam, displacing its plane, whereupon the

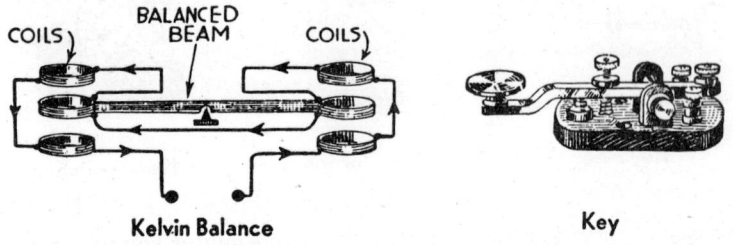

Kelvin Balance   Key

second Nicol prism shuts off an amount of light proportional to the rotation and to the signal voltage.
kathode.—A *cathode*.
kc.—Abbreviation for *kilocycles* (per second).
keeper.—A piece of iron placed between the poles of a *permanent magnet* to prevent loss in magnetic strength.
Kelvin balance.—An arrangement of coils in which flow of current produces fields which react with springs or weights for measurement of e.m.f. applied to the system. *See illustration.*
Kelvin temperature.—*Absolute temperature*. Equal to the centigrade temperature plus 273.1. Abbreviation is *K*.
Kennelly-Heaviside layer.—The *Heaviside layer*.
kenotron.—A high-vacuum hot-cathode rectifier tube capable of handling high voltages.

KERR CELL

Kerr cell.—A device in which the plane of a beam of *plane polarized light* may be rotated or tilted proportionately to a voltage applied on the cell. The light beam is passed transversely through a space between two conductive plates and through a quantity of nitrobenzene contained between the plates. The controlling voltage is applied across the two plates.

Kerr effects.—(*a*) *Double refraction* of a beam of *polarized light* when passed through an electrostatic field. (*b*) Rotation of a beam of *polarized light* which is reflected from the surface of a magnet.

key.—1—Same as gate. 2—A lever-operated switch for opening or closing electric circuits.

key chirp.—A sound produced in a receiver when the circuits of a telegraph transmitter continue to oscillate for a short time after the key has been opened.

key click.—*Key thump.*

key filter.—A combination of inductance, capacity and resistance

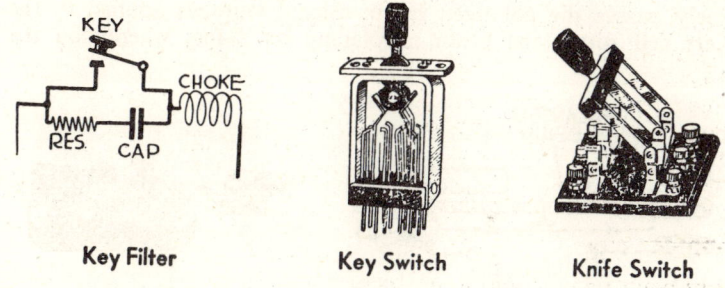

Key Filter   **Key Switch**   **Knife Switch**

or any of these in series or shunt with a transmitter key to prevent too sudden building up of voltages and currents in the controlled circuits. *See illustration.*

key modulation.—Variation of amplitude or frequency of a *carrier wave* by operation of a key in a transmitter circuit to form code signals.

key station.—A station at which a broadcast program originates and from which it is sent over wire lines to a number of other transmitters included in *chain broadcasting.*

key switch.—A switch operated by means of a cam and a small extension handle. *See illustration.*

key thump.—A sound produced in a receiver when a telegraph transmitter commences *oscillation* with great intensity upon closing of its key, also when there is excessive sparking at the key contacts.

keyed automatic gain control.—An automatic gain control system with which control voltage is secured from a tube allowed to conduct pulses of plate current only while its control grid or screen grid circuit is acted upon by voltage derived from horizontal sync pulses or from equivalent pulses taken from horizontal sweep or deflection circuits. Plate current pulses are filtered and smoothed to provide negative potential for grids of controlled amplifiers.

kilo-.—A prefix meaning one thousand times the unit. Abbreviated *k*.

kilocycle.—One thousand cycles per second, a unit of *frequency*. Abbreviated *kc*.

kilogram.—A unit of weight in the metric system. Equal to 2.205 pounds.

kilohertz.—One thousand cycles; one *kilocycle*.

kiloline.—A unit of *magnetic flux* equal to 1,000 magnetic lines of force.

kilometer.—A unit of length in the metric system of measurements. Equal to 0.6214 mile.

kilovolt-ampere.—A unit of measurement for *apparent power* in an alternating current circuit. The product of the number of amperes and the number of thousands of volts. Abbreviated *kva*.

kilowatt.—1,000 watts of electrical *power*. Abbreviated *kw*.

kilowatt-hour.—A unit of electrical *power* equal to 1,000 *watt-hours*.

Kinescope.—A trade name for television picture tubes made by Radio Corporation of America (RCA).

kinetic energy.—The energy or working ability which is a result of motion. Energy stored in a moving thing, an electric current, a moving field, etc., is called electro-kinetic energy. Compare *potential energy*.

Kirchoff's laws.—1—The sum of all the currents flowing toward any point or junction in an electric circuit is equal to the sum of all the currents flowing away from that point. 2—In any closed electric path the sum of all the impressed *e.m.fs.* is equal to the sum of all the *IR drops* of potential drops around the path.

knife switch.—A switch in which one contact is formed by a flat metal blade and the other by a piece or pieces of metal with which the blade makes contact. *See illustration*.

kw.—Abbreviation for *kilowatt*.

L

L.—Symbol for *self-inductance* in henrys.
l.—Symbol for *length*.
labyrinth enclosure.—An acoustic labyrinth.
lag.—*Angle of lag.*
lagging current.—A current in which maximum and zero amplitudes occur after the corresponding amplitudes of the applied voltage. The current in an *inductive circuit*. See illustration.
lagging phase.—Descriptive of an alternating quantity of which the zero point occurs after the zero point of another alternating quantity in the same circuit.
lambda (λ).—Greek letter symbol for *wavelength* in meters.

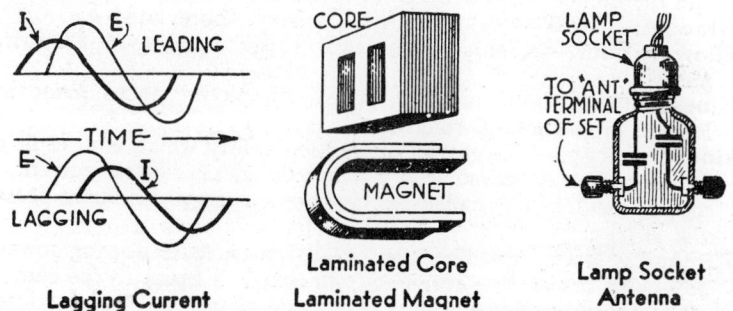

Lagging Current   Laminated Core / Laminated Magnet   Lamp Socket Antenna

lambert.—A unit of *brightness*. The brightness of a surface which is emitting or reflecting one *lumen* per square centimeter.
laminated core.—A magnetic core composed of many thin sheets of iron laid one over another, individually insulated by coatings of scale or varnish. Reduces the formation of *eddy currents*. See illustration.
laminated insulation.—Insulation made up of many thin layers.
laminated magnet.—A magnet having a *core* made up of several sheets or rods. See illustration.
lamp cord.—A flexible cord having a conductor of suitable size for use with incandescent lamps.
lamp socket antenna.—A connection made through a *fixed condenser* to one of the wires entering a lamp socket for the purpose of receiving radio waves picked up by the electric wiring. See illustration.
land line.—A transmission line carried over land areas.

L-ANTENNA

L-antenna.—An elevated horizontal aerial having the down lead connected at one end; having the form of an inverted "L". *See illustration.*

laryngophone.—A device attached to the outside of a person's throat and varying the resistance of an electric circuit in the manner of a *microphone* through vibrations of the flesh.

lateral vibration.—*Flexural vibration.* Compare with *transverse vibration.*

lattice.—The geometrical arrangement of atoms in a crystal. Sometimes called space lattice.

Lawrence color tube.—A type of single-gun color television picture tube.

layer winding.—A coil winding in which the adjacent turns are laid evenly side by side along the length of the winding. The entire winding may consist of one or more layers. *See illustration.*

lead.—1—The *angle of lead.* 2—An insulated conductor attached to an electrical device.

L-antenna

Layer Winding

leader tape.—Non magnetic tape used for splicing between lengths of, or for extending the ends of a recorded sound tape.

lead-in.—The conductor connecting the aerial of an antenna to the circuits in a transmitter or receiver. A down lead.

lead-in-bushing.—A tube of porcelain or other insulating material in which a *lead-in* is carried through walls or partitions. *See illustration.*

lead-in groove.—On a phonograph disc, the portion of the groove that guides the needle from the outer edge to the beginning of the sound track.

leading current.—A current in which maximum and zero amplitudes occur before the corresponding amplitudes of the applied voltage. The current in a *capacitive circuit.* Compare *lagging current.*

leading phase.—Descriptive of one alternating quantity of which the zero point occurs before the zero point of another alternating quantity in the same circuit.

leakage coefficient.—The ratio of the total magnetic flux to the flux which is useful in producing coupling. See *leakage flux.*

leakage current.—1—The current resulting from *volume leakage* or *surface leakage* in insulators. 2—The *dark current* of a photocell. 3—Current resulting from *condenser leakage.*

leakage detector.—A *ground detector*.
leakage flux.—The lines of force in a magnetic field which do not

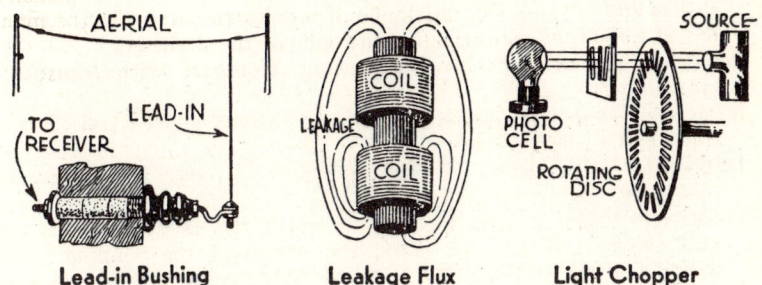

Lead-in Bushing   **Leakage Flux**   **Light Chopper**

encircle all of the turns in a coil and which are not effective in producing induction and *coupling*. *See illustration.*

leakage inductance.—That value of *self-inductance* of which the inductive reactance would equal the *leakage reactance*.

leakage reactance.—The number of ohms representing the energy loss which is due to failure of some of the lines of force from one winding to link with another winding to which the first is **coupled**. The loss due to deficient coupling.

leakage resistance.—The resistance to direct current of a condenser's *dielectric*.

leakance.—The conductance which allows flow of *leakage current*. A name sometimes applied to *dielectric conductance*.

Lecher wires.—A device for measurement of short *wavelengths*. Two long parallel wires are connected to a loop which is coupled to the output of an oscillator, *sta..ling waves* being produced on the wires. Movement of a short-circuiting bridge along the wires will allow resonance with the oscillator frequency to be indicated at certain points. The length of the wires will be equal to an odd number of quarter-wavelengths.

Lenard rays.—*Cathode rays* which are driven through a specially prepared "window" of a *cathode ray tube* into the space outside the tube where they may be applied to various materials.

lens.—A piece of transparent material, as glass, which is so shaped as to cause *refraction* and change the direction of rays of light passing through it.

lens axis.—The *optical axis*.

Lenz's laws.—1—An e.m.f. which is produced from a magnetic field by induction tends to set up a current which has a second field opposing any change in the original magnetic field. 2—The total number of cuttings per second between the magnetic lines and the conductors in which induction produces an e.m.f., divided by 100,000,000, equals the number of induced volts.

level indicator.—A meter or other means for visually indicating audio voltage or power, to guard against overloading during tape, wire, or phonograph recording.

level of signal.—A measure of modulation strength at various amplitudes of a composite television signal, with reference to zero amplitude of the carrier. Peaks of sync pulses are at 100 per cent, the black level is at approximately 75 per cent, while picture signals extend from darkest at the black level to about 15 per cent for brightest. Compare negative transmission.

Leyden jar.—A glass vessel having a thin layer of metal on the outer surface and a similar layer on the inner surface, thus forming a *condenser* with the glass as the dielectric and the metal as the plates.

l-f.—Abbreviation for *low frequency.*

life test.—Operation of a device under conditions approximating normal use for a period of time sufficient to determine changes which would occur in actual service.

light.—Visible light is radiation at wavelengths which affect the sense of sight; wavelengths lying approximately between 0.000039 and 0.000076 centimeters, or between 3900 and 7600 *Angstrom units,* from violet to red. The word also is used to mean *luminous flux,* and in some cases to mean any form of radiation or *radiant energy.*

light cell.—A *photoelectric cell.*

light chopper.—A device for interrupting the light passing to a photocell, thus producing a pulsating current required for *carrier frequency amplification. See illustration.*

light flux.—*Luminous flux.*

light socket antenna.—A *lamp socket antenna.*

light spectrum.—The *visible spectrum.*

light unit.—A unit of *luminous intensity* equal to about 9.615 candlepower. A *carcel.*

light valve.—1—A device utilizing variations in frequency and intensity of current resulting from sounds to vary the width of an opening through which passes the light falling on a photoelectric cell. 2—A *Karolus light valve.*

light waves.—Waves of radiant energy at frequencies which affect the sense of sight, the waves consisting of vibrations taking place transversely to the lines of propagation but otherwise having no definite direction of motion. See *light.*

lightning arrester.—A device which allows current surges in an *antenna circuit* to pass directly to ground without flowing through the receiver circuits. A small gap between two electrodes in air or a vacuum, connected between aerial and ground. Sometimes several small gaps in series between antenna and ground. *See illustration*

lightning switch.—An *antenna switch.*

LIMITER

limiter.—1—In a frequency-modulation receiver, a tube operated to reach plate current cutoff or saturation on input signals exceeding a certain amplitude, thereby preventing passage to the demodulator of amplitude modulation exceeding this value. 2—A sync limiter.

line.—1—A metallic conductor in a circuit between devices at a distance from each other. 2—Line of force; see *magnetic lines of force* and *electrostatic lines of force*. 3—Line of induction; lines which represent the *magnetic flux* in a magnetic circuit. In air, having unit permeability, the number of lines of induction is the same as the number of *magnetic lines of force*.

line amplifier.—An *audio frequency amplifier* taking the output of one or more microphones and, after amplification, feeding the signals to a transmission line.

line drop.—The electromotive force used in sending current through the conductors between a source and its load. Measured in *volts*. See illustration.

line frequency.—The frequency at which occur horizontal lines in television pictures. For black-and-white or monochrome this frequency is 15,750 cycles per second, and for color television is 15734.264 cycles per second.

line loss.—*Attenuation* occurring in a transmission line and in devices forming a part of the line.

line radio.—*Wired radio*.

line relay.—A *relay* having its coil connected in a main line.

line voltage regulator.—A device which reduces the effects of changes in supply line voltage and applies a fairly uniform voltage to a connected circuit, such as to the power unit for a receiver. A *ballast tube* or any type of *voltage regulator*.

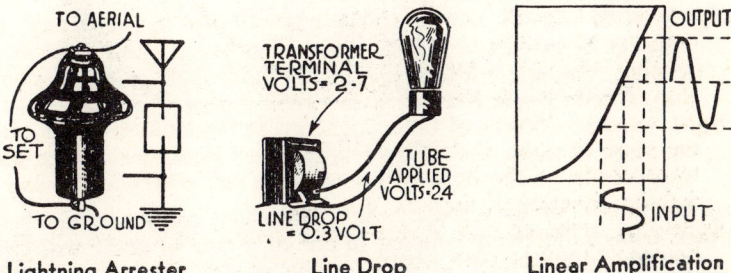

Lightning Arrester   Line Drop   Linear Amplification

linear.—A relation between electrical quantities such that change in one is accompanied by an exactly proportional change in another.

linear amplification.—Amplification in which signal output voltage at any frequency and any value is directly proportional to the input voltage. *See illustration.*

linear constant.—An electrical constant measured per unit length of transmission line.

linear decrement.—The difference between successive amplitudes in the same polarity of a *damped oscillation or wave* in which these differences are of the same or constant value.

linear detection.—Any form of detection which results in output voltages directly proportional to the applied signal voltages throughout the useful range of the system.

linear distortion.—Amplitude distortion.

linearity.—A characteristic of television receiver pictures which relates to shape, relative positions and sizes of objects and lines in the reproduction as related to these features in the original scene and their representation in received signals. There is linearity when objects and outlines are not distorted, otherwise there is non-linearity.

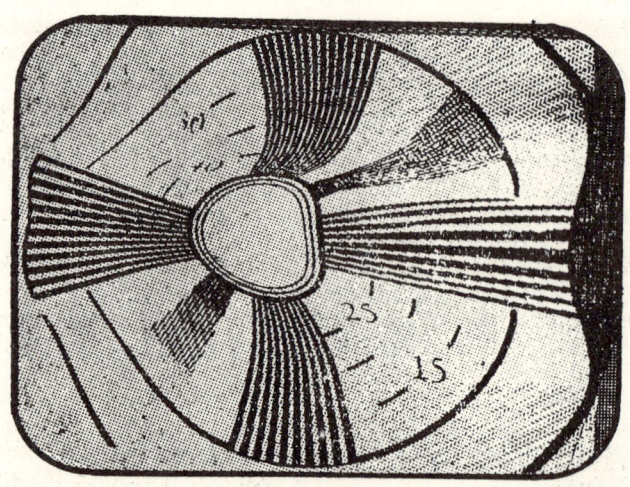

Linearity (Faulty)

link circuit.—Two *inductance coils*, conductively connected together, each being inductively coupled to one of a pair of additional circuits between which a transfer of energy is desired. *See illustration.*

link coupling.—Coupling provided between two circuits by an intermediate *link circuit*.

link fuse.—A *fuse* consisting of a short length of open wire between two fastenings.

linkage.—*Flux linkage.*

LISSAJOUS FIGURES

Lissajous figures.—Curves which are the result of combining two *simple harmonic motions*, such as the motions of two pendulums differing in phase.

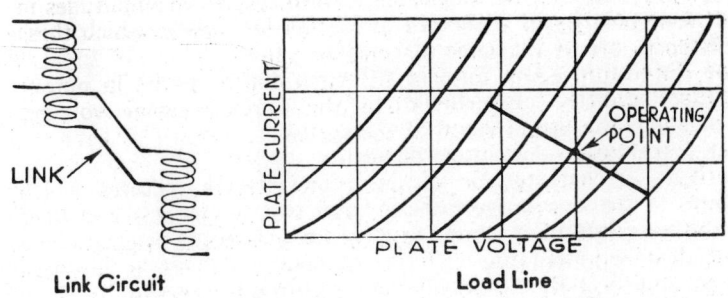

Link Circuit   Load Line

litzendraht wire.—A conductor formed of numerous separately enamelled strands which are transposed at intervals so that each one takes up successively all possible positions in the cross section of the whole conductor. Used to reduce *skin effect*. The name often is abbreviated to "litz".

load.—A *resistance, reactance, impedance,* or combination of these in which work is done by electric power.

load center.—A point in a circuit at which, for purposes of calculation, it may be assumed that the whole load is located.

load line.—A line drawn upon a series of *plate characteristics* and used for determination of plate current change with grid voltage change, also for determination of *harmonic distortion*, when a specified plate load is used. The load line crosses the intersection of the grid bias curve with the plate voltage ordinate, and the slope of the line is made proportional to the reciprocal of the plate load in ohms. The ratio of milliamperes plate current change to volts change in plate potential, as shown by the load line slope, is made equal to the ratio of 1000 to the load resistance in ohms. *See illustration.*

loaded antenna.—An antenna circuit containing a *loading coil*.

loaded line.—A circuit in which the normal reactance has been altered by addition of *lumped reactances* to improve the performance.

loading coil.—A coil inserted in a circuit for the purpose of increasing the total inductance without providing additional coupling with any other circuit. Used to lower the *resonance frequency* of the circuit.

lobe.—One of the curved outlines on an antenna polar diagram showing signal pickup ability in various directions as a percentage of maximum pickup.

local amplifier.—An *audio frequency amplifier* placed near one or more loud speakers, to increase the strength of a transmission line signal to the degree required for speaker operation.

local oscillations.—1—Any *oscillatory current* generated within apparatus at the receiving end of a radio transmission system. 2—Oscillatory currents produced in certain receivers, as superheterodynes, or in frequency meters; and combined with other oscillations received from external sources to generate *beat frequencies* or create the condition of *zero beat*.

lock-in base.—A tube base at whose center is an extension with ball-shaped tip that snaps between springs in a socket, to hold the tube firmly in place.

log.—A record of tuning dial settings. A record of stations with which a transmitter has been in communication.

logarithm.—The power to which a number, called the base, must be raised to equal another number called the antilogarithm. The base for common logarithms or Briggs logarithms is 10; the base for natural logarithms is 2.71828.

logarithmic decrement.—A quantity which is a measure of the *damping* in an oscillatory circuit, and which is inversely proportional to the *selectivity*. It is the natural logarithm of the ratio of two successive maximum amplitudes of the same polarity. In a simple radio circuit the logarithmic decrement is equal to the square root of the quotient of the capacity divided by the inductance, multiplied by π (3.1416) times the resistance. The symbol is the Greek letter delta (δ). *See illustration*.

logarithmic horn.—An *exponential horn*.

long waves.—Radio waves having a length of 600 meters or more; frequencies lower than 500 kilocycles.

longitudinal effect.—Electrical polarization and mechanical extension in the same direction in a *piezo-electric crystal*.

longitudinal magnetization.—Producing magnetized areas on a recording tape along the direction in which the tape moves.

loom.—*Circular loom*.

loop.—1—An *antinode*. 2—A *loop antenna*.

loop antenna.—A radio *antenna* consisting of one or more turns or loops of conductor supported on an open framework which is approximately rectangular or circular. A tuning condenser is connected across the ends of the loop conductor, the combination being made resonant at the received or transmitted frequency. *See illustration*.

loop circuit.—A circuit in which all energy consuming devices are at the same electrical distance, through conductors, from the source.

loop receiver.—A receiver especially designed for operation with a *loop antenna*.

loose coupler.—An inductive coupling device providing a small *coupling coefficient*.

longitudinal vibration.—Motion of the disturbed particles forming a wave in the direction of wave propagation. A characteristic of sound waves. Compare *transverse vibration*.

loose coupling.—A coupling providing little transfer of energy;

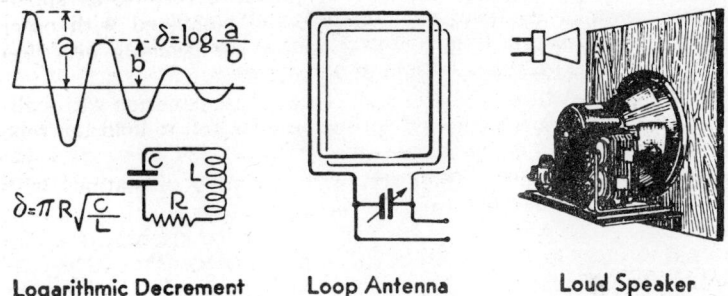

Logarithmic Decrement   Loop Antenna   Loud Speaker

usually one in which the *coupling coefficient* is less than 0.5. A coupling between two circuits in which the *mutual inductance* is small compared with the *self-inductance*.

loud speaker.—A device which produces sound waves from electric power applied to it, the sounds being of sufficient intensity to provide good audibility in a room or the open air. *See illustration*.

low band.—The very-high frequency television broadcast channels numbers 2 to 6 inclusive.

low frequency.—By definition, a frequency in the range from 30 to 300 kilocycles per second.

low frequency compensation.—Means for extending the gain of a broad band amplifier to low audio or video frequencies, usually by employing suitable relative values of blocking or coupling capacitors, grid return resistors, and plate load decoupling elements.

low-loss.—Having but little *high frequency resistance*, consequently having but little loss of radio frequency energy.

low pass filter.—A filter system which is designed to allow comparatively free passage of low frequency and direct currents in a circuit but which greatly attenuates all frequencies above a certain *cutoff frequency*. *See illustration*.

low voltage winding.—In a transformer, the winding having fewer turns. *See illustration*.

lower side band.—The *side frequency* which is less than the *carrier frequency*.

lumen.—A unit of light. That *luminous flux* from a point source of one *international candle* which passes through a solid unit angle, or passes through one square centimeter of a surface at a distance of one centimeter from the source. Equal approximately to the total flux from such a source divided by 12.57. The light reaching an area of one square foot when the illumination intensity is one *foot-candle*.

lumen-hour.—A unit of quantity of light, equal to a luminous flux of one *lumen* flowing for one hour.

luminance.—A word whose meaning is generally equivalent to brightness in describing the qualities of light or illumination. Often measured in candlepower per unit of surface area.

luminance channel.—Color television circuit primarily intended to carry luminance signals, although other signals may travel in part of the same path.

luminance signal.—A color television signal which is the equivalent of a video signal in black-and-white television. The signal which represents changes of brightness, also lights and shadows or fine details, but does not control coloring.

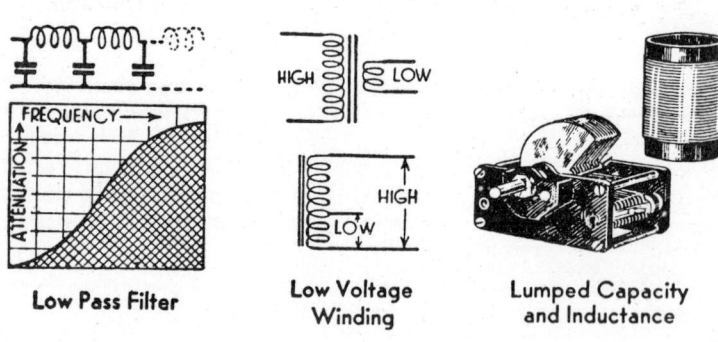

Low Pass Filter   Low Voltage Winding   Lumped Capacity and Inductance

luminescence.—An emission of light not due to heating, but resulting from electrical action, chemical action, etc. Compare *incandescence*.

luminosity.—The quality of being *luminous*.

luminosity curve.—A curve showing the relation between *wavelength* of light and the *luminous flux*.

luminous.—Descriptive of a body which is emitting *light*.

luminous efficiency.—The ratio of the *luminous flux* to the *radiant flux* in the radiation from a given source. The unit is lumens per watt.

luminous flux.—1—The rate of flow of light radiation with respect to the sense of sight. Measured in *lumens*. The symbol is F or f. 2—Sometimes used as meaning quantity of light, the product of flux and time.

luminous intensity.—The amount of light per unit of area of the emitting source. The *luminous flux* emitted by a source through a solid unit angle. The unit is the *international candle*. The symbol is I.

LUMPED CAPACITY

lumped capacity.—A capacity existing in a relatively small physical space, such as in a condenser of usual construction. *See illustration.*

lumped constant. — Capacitance or inductance concentrated in capacitors or coiled inductors, as distinguished from capacitance and inductance of tubes, wiring, and other circuit parts.

lumped inductance.—An *inductance* existing in coil windings.

lumped plate voltage.—The plate voltage plus the product of amplification coefficient and grid voltage on the tube.

lumped reactance.—A condenser having capacity or a coil having inductance concentrated within small space.

lux.—A unit of illumination equal to one lumen per square meter. Equal to 0.1 *milliphot* and approximately to 0.093 *foot-candle.*

M

M.—A symbol for megohms of resistance.

M.—Symbol for *mutual inductance* in henrys. Abbreviation for the prefix *mega-*.

m.—Symbol for *meter*, a measure of length. Abbreviation for the prefix *milli-*.

magic eye.—An electron-ray tube.

magnet.—A piece of iron or steel which has the property of attracting other pieces of iron and steel. A *permanent magnet* or a temporary magnet. Compare *electromagnet*.

magnet wire.—Insulated copper wire of gage sizes generally used for windings of electromagnets, choke coils, transformers, etc.

magnetic.—Having the properties of a *magnet:* utilizing magnetic lines of force.

magnetic bias.—In tape or wire recording a high-frequency magnetizing current which is combined with audio frequency currents for recorded sounds, so that magnetic flux in recording elements varies only along a relatively straight portion of the magnetization curve. This avoids distortion which would result from operation on a bend of the curve.

magnetic circuit.—A completely closed path through which pass *magnetic lines of force* or magnetic flux. With a magnet; the circuit consists both of the path within the magnet and the path external to the magnet. *See illustration.*

magnetic compass.—A device which indicates geographical direction by action of the earth's magnetic field on a suspended bar magnet.

magnetic component.—The *electromagnetic component*.

magnetic convergence.—Convergence of the electron beams in a three-gun color television picture tube by means of magnetic fields from coils or permanent magnets mounted around the outside of the tube neck.

magnetic core.—See *core*.

magnetic coupling.—*Inductive coupling*.

magnetic current.—The rate of change of *magnetic flux*.

magnetic curve.—The *magnetization curve*.

magnetic cycle.—The series of changes which occur in the magnetization of a material being affected by an alternating *magnetomotive force*.

magnetic damping.—*Damping* of instruments by means of the reaction between two magnetic fields acting to make the pointer dead beat in its movement. The magnetic fields are produced from

MAGNETIC DEFLECTION

eddy currents set up in a piece of metal attached to the pointer and free to move between the poles of the instrument's permanent magnet.

magnetic deflection.—Deflection or bending of the electron beam in a television picture tube or cathode-ray tube by magnetic fields produced by sawtooth currents in pairs of coils located in the yoke around the picture tube neck. The field between one pair of coils whose axes are in line causes horizontal deflection, while the field between another pair causes vertical deflection.

magnetic energy.—*Electrokinetic energy.*

magnetic fatigue.—An increase of *hysteresis* occurring after many changes of magnetism. *Aging* of iron.

magnetic field.—The space occupied by magnetic lines of force. The space within which there are magnetic effects. *See illustration.*

magnetic field intensity.—*Magnetizing force.*

magnetic flux.—The total flow of *magnetic lines of force* or magnet-

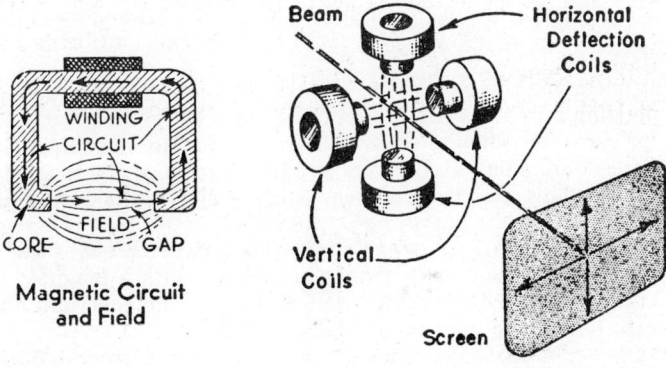

Magnetic Circuit and Field

Magnetic Deflection

ism through a magnetic circuit. Measured in *maxwells* or in number of magnetic lines. The number of lines is equal to the *magnetomotive force* in gilberts divided by the *reluctance* in oersteds. The symbol is the Greek letter phi (ϕ).

magnetic flux density.—The number of *magnetic lines of force* in a unit cross sectional area of a magnetic field. Measured in *gausses* or in lines per square inch. The symbol is B.

magnetic focusing.—Focusing the electron beam in a television picture tube or cathode-ray tube by means of fields from permanent magnets or from an electro-magnet mounted outside the tube neck. Internal magnets are used in some designs.

magnetic friction.—The retarding force exerted on a *magnetic material* moving in a *magnetic field.*

MAGNETIC HYSTERESIS

magnetic hysteresis.—*Hysteresis*.
magnetic induction.—1—The effect of a magnetic field by which it makes a piece of iron or steel into a magnet. *See illustration*. 2—*Magnetic flux density*.
magnetic inductive capacity.—*Permeability*.
magnetic instrument.—An instrument which indicates the current in a circuit by using the reaction between magnetic fields which are produced or controlled by the measured current. *Moving coil instruments, vane type instruments*, etc.
magnetic intensity.—*Magnetic flux density*.
magnetic interrupter.—Two contacts arranged to suddenly open and close an electric circuit. The contacts are automatically operated by an *electromagnet* connected in the circuit, opening of the contacts stopping the magnetizing current and allowing the contacts to again close.
magnetic key.—A *key switch* in which the current carrying contacts are held together by an electromagnet and are allowed to open only at an instant of zero current after the key is released.
magnetic lag.—*Hysteresis*.
magnetic leakage.—*Leakage flux*.
magnetic lines of force.—1—In the space around a current carrying conductor, or a magnet, lines which represent the direction in

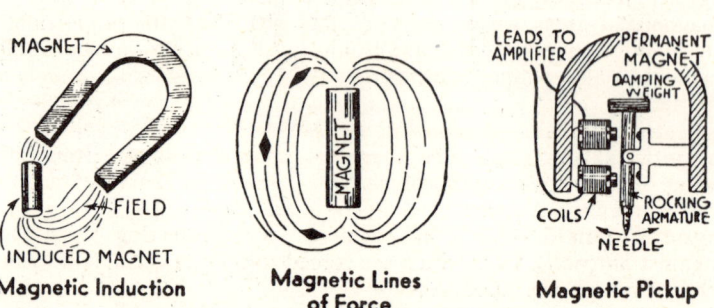

Magnetic Induction **Magnetic Lines of Force** **Magnetic Pickup**

which a short magnetized needle would point as moved through the space. Also called unit lines of force or G.G.S. lines of force. The *magnetizing force* at any point is indicated by the number of lines per unit of area. In a unit magnetic field there is one line per square centimeter of cross section. *See illustration*. 2—The magnetic line of force is a unit of *magnetic flux*. Lines per square centimeter or per square inch are measures of *magnetic flux density*.
magnetic linkage.—*Flux linkage*.
magnetic loud speaker.—Any *loud speaker* in which the sound producing mechanical movements are the result of magnetic reactions.
magnetic material.—Any substance which may become a *magnet*.

MAGNETIC MICROPHONE

Iron and steel are the chief magnetic materials. Nickel and cobalt are slightly magnetic.

magnetic microphone.—Any *microphone* with which the signal output is generated in a coil which is moved within a magnetic field. An *electromagnetic microphone.*

magnetic modulator.—A device which causes *modulation* by the unequal changes of magnetic flux in an iron core acted upon by a magnetizing current which varies at the modulating frequency.

magnetic moment.—The product of the strength of a magnetic pole and the distance between the poles.

magnetic needle.—A small permanent magnet pivoted or suspended so that it moves into line with the direction of a *magnetic field* surrounding it.

magnetic pickup.—A *phonograph pickup* in which the electrical output results from needle controlled movement of a portion of a magnetic field, the changing flux producing signal currents in a coil associated with the field. *See illustration.*

magnetic polarity.—See *polarity.*

magnetic pole.—The place at which *magnetic lines of force* leave or enter a magnet or source of magnetic flux. See *unit pole.*

magnetic potential.—The potential through which a *unit pole* is moved in bringing it from infinity (zero potential) to the point being considered.

magnetic recording.—Formation on a recording tape or wire of magnetized areas whose positions and strengths are proportional to sound frequencies and amplitudes. These areas induce audio currents and voltages when the tape or wire is run through a reproducer.

magnetic rectifier.—A *vibrating rectifier.*

magnetic saturation.—The magnetic condition of an iron core when further increase in *magnetizing current* in the winding produces but little increase of *magnetic flux.*

magnetic shield.—An iron enclosure for protecting apparatus against *magnetic fields* which are forced to flow in the shield rather than through the apparatus.

magnetic shunt.—A piece of iron located near field poles of a magnet and used to weaken or control the strength of the magnetism in the space between the poles. Often used to regulate the deflection of pointers in *magnetic instruments.* *See illustration, page following.*

magnetic spectrum.—The effect on a fluorescent screen of *cathode rays* which have passed through a magnetic field.

magnetic storm.—Irregular changes in the paths taken by the earth's magnetic field, movements of these lines of force causing *static* interference.

magnetic strength.—*Magnetic flux density.*

magnetic stress.—A *stress* due to the forces existing in a magnetic field.

magnetic tick.—A faint ticking sound heard upon suddenly magnetizing or demagnetizing iron. The Page effect.

magnetic vane instrument.—A *vane type instrument*.

magnetism.—The force which produces magnetic effects and which appears in a *magnet*.

magnetite.—An oxide of iron, sometimes found in a naturally magnetized state and called loadstone or lodestone.

magnetization.—The magnetic strength of a body. Also, the act of making magnetic or *magnetic induction*.

magnetization curve.—A curve showing the relation between *magnetizing force (H)* in ampere-turns or gilberts per centimeter and the *magnetic flux density (B)* in lines per square centimeter produced in a magnetic material. A B-H curve. *See illustration*.

magnetizing current.—The current which is required in magnetizing an iron *core*, a current just sufficient to produce a *counter-electromotive force* equal to the applied voltage. Generally used as meaning the current which flows in the primary of a transformer when the secondary is open circuited, this being equal to the above magnetizing current plus that required to overcome eddy current and hysteresis losses.

magnetizing force.—The *magnetomotive force* per unit length of magnetic circuit. The magnetic pressure or force required to produce a given number of flux lines in a certain length of a magnetic path. Measured in *gilberts* or *ampere-turns* per centimeter length

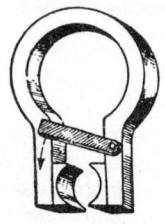
Magnetic Shunt

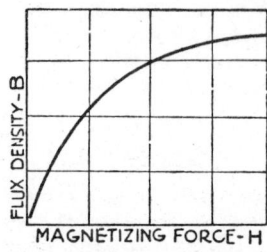
Magnetization Curve

Magnetostriction Oscillator

of path, or else in ampere-turns per inch length of path. The symbol is H.

magneto.—An electric generator using *permanent magnets* to produce its field.

magnetomotive force.—The force which acts to produce *magnetic flux* or lines of force in a magnetic circuit. Measured in *ampere-turns* or in *gilberts*. Magnetomotive force in gilberts is equal to the product of the magnetic flux in lines and the reluctance in oersteds. The symbol is F. Compare *magnetizing force*.

MAGNETOSTRICTION

magnetostriction.—The effect which produces a change in the shape or size of a body which is magnetized. The quality is possessed by nickel and some magnetic alloys.

magnetostriction loud speaker.—A *magnetic loud speaker* in which the mechanical forces producing sound waves are the result of *magnetostriction*.

magnetostriction oscillator.—A low frequency oscillator using a rod possessing magnetostriction qualities as a means for regenerative coupling in a vacuum tube circuit. The circuit generates oscillating currents controlled in frequency by the natural frequency at which the rod vibrates. This frequency is dependent on sound velocity in the rod's material and on the dimensions of the rod. The action is similar to the *piezo-electric effect*. See *illustration*.

magnetron.—A two-element vacuum tube or diode used as an *oscillator*. A varying magnetic field around a large filament carrying alternating current deflects the electron stream and causes variations in the electron flow to the plate, which forms a cylinder enclosing the filament.

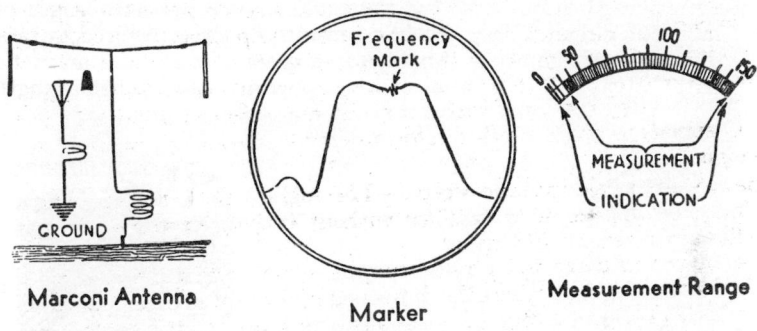

Marconi Antenna   Marker   Measurement Range

manganin.—A resistance metal composed of copper, manganese and nickel.

Mansbridge condenser.—A *fixed condenser* having metal foil for plates and waxed paper for dielectric. A paper condenser.

manual regulation.—Control or adjustment by hand operated devices

Marconi antenna.—An antenna circuit of which the ground forms an essential part. See *illustration*.

marker.—A small break, dip, or jog at the position of a certain frequency along a frequency response.

marker generator.—A signal generator for production of voltages at accurately known frequencies for application to amplifiers, receivers, or various other devices whose frequency response is being observed on an oscilloscope.

mask.—A frame around the opening in a television receiver cabinet through which appears the picture tube viewing screen.

masking effect.—The effect by which one sound which is increasing in intensity finally prevents hearing another sound of different pitch.

mass.—The quantity of matter contained in a body. The quality of a body by which it exhibits *inertia*, and which allows it to be acted upon by gravity. Measured in units of weight; pound, gram, etc.

mass resistivity.—The resistance of a certain mass or weight of a substance when it occupies a certain length. For example, the resistance of a wire of uniform cross section, weighing one gram and having a length of one meter.

mast.—An elevated support for a radio antenna.

master disc.—A *wax master* or a *metal master* in the disc method of sound recording.

master negative.—The *metal master* in disc recording.

matched impedance.—The condition under which there is an *impedance match* between two connected circuits.

matching pad.—A combination of resistors used on the free end of an instrument cable to provide on the cable side an impedance approximately equal to that of the cable, and on the other side an impedance approximately equal to input impedance of a receiver or other device to which connection is made.

matrix.—In a color television receiver, circuits wherein the luminance signal combines with color-difference signals or with I- and Q-signals to produce other color-difference signals or primary color signals.

maximum peak inverse volts.—The highest peak voltage which may be applied to a *rectifier* without causing excessive current flow in a reverse direction.

maximum undistorted power output.—The output power obtainable from a tube when the input signal does not exceed a value which produces a five per cent distortion due to the second and higher *harmonic* components which are introduced by the tube.

maxwell.—The C.G.S. unit of *magnetic flux*, equal to one magnetic line of force.

May effect.—The effect of light on electrical resistance, as observed in *selenium*.

mean effective value.—The *effective* value.

mean free path.—The average distance traveled in a gas by *electrons* between collision with atoms, or the distance traveled by atoms between collisions with other atoms. See *ionization*

mean spherical candlepower.—The average candlepower of a source in all directions. Equal approximately to 12.566 times the *luminous flux* in lumens.

measurement range.—The range of values within which an instrument's requirements for accuracy are satisfied. Compare *indication range. See illustration.*

mechanical axis.—A *Y-axis* of a quartz crystal.

mechanical impedance.—The opposition of a mechanical system to motion; determining the velocity of motion which results from a certain applied alternating mechanical force. Equal to the applied force divided by the resulting velocity in the same direction, measured at the point of application. One unit of mechanical impedance exists when a force of one dyne produces a velocity of one centimeter per second. A similar unit is used for measurement of *mechanical resistance* and *mechanical reactance*.

mechanical reactance.—That portion of the *mechanical impedance* which results from the *compliance* of the mechanical part. The portion of the mechanical impedance which is analogous to the reactance in an alternating current circuit.

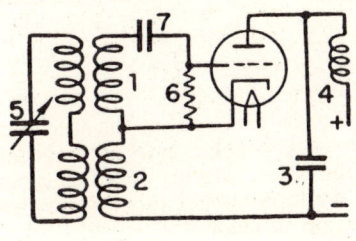

Meissner Oscillator

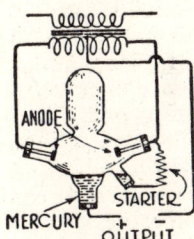

Mercury Arc Rectifier

mechanical resistance.—The result found by dividing the power absorbed in a mechanical system by the square of the alternating velocity at the point where force is applied. The component of the *mechanical impedance* corresponding to ohmic resistance in an alternating current circuit.

medium frequency.—By definition, a frequency in the range from 300 to 3000 kilocycles per second.

meg- or mega-.—A prefix meaning one million times the unit. Abbreviated M.

megger.—An *ohmmeter* for measuring high resistances. A magneto sends current through two coil systems, one containing the resistance to be measured in its circuit. The relative torque of the two systems determines the position of an indicating pointer.

Meissner oscillator.—A *radio frequency oscillator* in which a winding in the tube plate circuit and another in the grid circuit are both coupled to a third winding or to a *link circuit* which is tuned to the desired frequency. Feedback takes place from plate circuit to grid circuit through the link. *See illustration.*

mercury arc rectifier.—A *gaseous conduction rectifier* in which primary electron emission takes place from a heated spot in a pool of mercury, the electrons producing ionization and a heavy current in a vapor of mercury between the heated spot and an anode. Rectified current passes from either of two anodes to the mercury.

mercury condenser.—A *variable condenser* with air dielectric, the active plate area being varied by moving the plates into and out of a mercury bath.

mercury interrupter.—A device for rapidly opening and closing a circuit which is carried through a jet of mercury striking against a toothed wheel or a metallic grid which completes the circuit.

mercury switch.—An electric switch consisting of two metallic terminals passed through the walls of an insulating tube containing a small amount of mercury. With the tube level, the mercury connects the two terminals and with the tube tilted the mercury flows to one end and breaks the contact.

mercury vapor rectifier.—A *gaseous conduction rectifier* having a heated cathode and a cold plate or anode in a gas consisting of mercury vapor at low pressure. The cathode may be of the filament type or of the heater type. *See illustration.*

Mershon condenser.—The trade name for an *electrolytic condenser* using a liquid electrolyte.

metal backed screen.—An aluminized screen. Sometimes a silver screen.

metal tube.—An electronic tube whose evacuated envelope is metal. In the envelope are insulation beads through which pass leads for internal elements. Also a television picture tube whose cone or flare is of metal, with the remainder of glass.

metallic line or circuit.—An electric circuit with both sides carried by wires, neither side being grounded.

meter.—1—A device which indicates a value of a measured electrical quantity. 2—A unit of length in the metric system of measurements; equal to 39.37 inches, 3.281 feet, 1.094 yards. Abbreviated m.

meter-amperes.—A measure of the *radiation* strength of a radio transmitter. The number of meter-amperes is equal to the maximum antenna current in amperes multiplied by the antenna effective height in meters.

meters per millimeter.—In a *piezo-electric resonator*, the fundamental wavelength in meters divided by the dimension (in millimeters) along which vibrations take place.

metric system.—A system of units based on the meter for length and the gram for mass with decimal divisions and multiples of these for the derived units.

mf or mfd.—Abbreviation for microfarad of capacitance.

mho.—A unit of *conductivity* or of *admittance*. The reciprocal of one *ohm*.

mica.—A mineral occurring in crystalline laminations or layers. One of the best insulators and condenser dielectrics. Dielectric constant 3.5 to 6.0.

MICA CONDENSER

mica condenser.—A condenser using mica as its dielectric. *See illustration*.

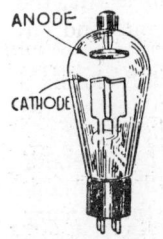
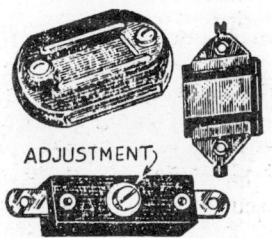
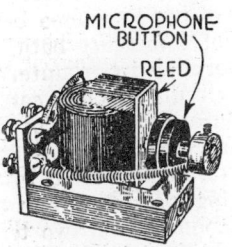

Mercury Vapor Rectifier   Mica Condenser   Microphone Hummer

micabond.—Sheet insulation consisting of mica and shellac, or of mica laminated with paper or cloth.

micanite.—Insulation made by holding together small pieces of mica with a binder cement.

micarta.—An insulating composition made of paper and mica.

micro-.—A prefix meaning one-millionth of the unit. The symbol for micro- is the Greek letter mu (μ).

microfarad.—A unit of electrostatic capacity equal to the one-millionth of a *farad*. Abbreviated *mfd*.

microhenry.—A unit of *inductance* equal to the one-millionth part of a *henry*.

microhm.—A unit of *resistance* equal to the one-millionth part of an *ohm*.

micromho.—A unit of *conductivity* or of *conductance* equal to the one-millionth part of a *mho*. The usual unit for measuring *mutual conductance* of a vacuum tube.

micromicro-.—A prefix meaning one-millionth part of the millionth of a unit.

micron.—1—The one-millionth part of a meter. A unit of *wavelength*. See *millimicron*. 2—A unit of pressure equal to that exerted by a column of mercury 0.001 millimeter in height.

microphone.—A device for converting *sound waves* into corresponding changes of electric power. An instrument in which an impedance formed by inductance, capacity, resistance or a combination of these is altered in value by changes in air pressure which result from sounds.

microphone amplifier.—1—An *audio frequency amplifier* placed close to a microphone and amplifying the microphone output to a value suitable for line transmission or for application to a station amplifier. 2—An *audio frequency amplifier* in which variations of signal current passing through a magnet winding cause the magnet to vary the pressure on a carbon contact, thus varying the resistance of a second circuit containing the carbon.

microphone button.—A mass of loosely held carbon particles of which the resistance is altered by variations of pressure on the housing.

microphone hiss.—A hissing sound existing in the amplified audio frequency output of a *carbon microphone*, the result of current flow between particles of carbon in the microphone.

microphone hummer.—A device making use of a vibrating electromagnetically operated reed and a *microphone button* for the production of alternating currents at audio frequencies. See *illustration*.

microphone response.—The ratio of the signal voltage applied from a *microphone* on a connected tube control grid to the pressure in *bars* on the microphone diaphragm.

microphone singing.—A sustained audible note from a loud speaker whose sound waves reach the microphone connected to the input of the amplifying system feeding the loud speaker. See *acoustic feedback*.

microphone transformer.—A transformer having impedance suitable for coupling a *microphone* to a tube, a line, a volume control, an amplifier or other device.

microphone transmitter.—A *microphone*.

microphonic.—Descriptive of a device, such as an amplifying tube, in which mechanical movement of internal elements causes a corresponding variation of circuit impedance. If the mechanical movements are produced originally by sound waves from a loud speaker, the result is a sustained audible note.

microphonic contact.—An imperfect joint between two conductors such that slight relative movement between them causes a corresponding change in *resistance*.

microvolter.—An instrument for generating measured values of radio frequency voltages at very low potentials.

microvolts or millivolts per meter.—A measure of the *radio field intensity* at a given point of reception. Equal to the number of microvolts or millivolts potential developed between the aerial and ground of an antenna system, divided by the *effective height of the antenna* in meters. See *illustration*.

mike.—A *microphone*.

mil.—One one-thousandth of an inch of length.

mil-foot.—A unit of conductor size; equal to a cross sectional area of one *circular mil* and a length of one foot.

Miller effect.—A change of input capacitance in an amplifier tube when the grid bias voltage is altered. Such changes occur during operation of either automatic or manual gain controls or volume controls.

millimeter.—A measure of length in the metric system of measurements. Equal to 0.03937 inch. Abbreviated *mm*.

millimicro-.—A prefix meaning the thousandth part of a millionth of the unit.

millimicron.—A unit of *wavelength* equal to the one-billionth part of a meter, or to ten *Angstrom units*. The symbol is $m\mu$.

milliphot.—A unit of *illumination* equal to one-thousandth of a lumen per square centimeter. Equal to 10 *lux* or approximately to 0.93 *foot-candle*. See *phot*.

millivolt.—The one-thousandth part of a volt. Abbreviated *mv*.

millivoltmeter.—A voltmeter calibrated to read directly in millivolts or in thousandths of a volt.

millivolts per meter.—See *microvolts or millivolts per meter*.

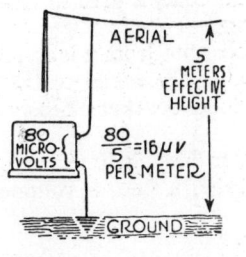

Microvolts per Meter

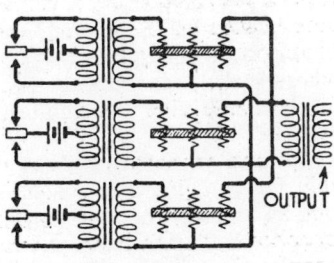

Mixer

minus color signals.—In color television, a chrominance signal representing a color which is the complementary of a plus color signal and whose phase differs by 180 degrees from that of the plus signal. Were plus signals yellow, red and magenta, the respective minus complementaries would be blue, cyan and green.

mirror backed screen.—An aluminized screen.

mixed highs.—A color television luminance signal formed by combining certain proportions of the three primary color signals to form the equivalent of a signal for black-and-white or monochrome color television.

mixer.—A control device which allows varying of the intensities of signal voltage from several different microphones or other sources which are combined or blended in various proportions to form the input to an amplifier. *See illustration*.

mixer tube.—In superheterodyne reception, a tube in which beat together a carrier frequency and a frequency from an oscillator which part of the tuner or receiver. The beat frequency, which is equal to the difference between carrier and oscillator frequencies, becomes the intermediate frequency which carries the same signal modulation as the carrier.

mm.—Abbreviation for millimeter of length.

mmf or mmfd.—Abbreviation for micromicrofarad of capacitance.

mobile receiver.—A radio receiver designed to function while in motion. A receiver for use in an airplane, an automobile, a ship, etc.

modulated oscillator.—A radio frequency oscillator of which the output carries *modulation* at an audio frequency.

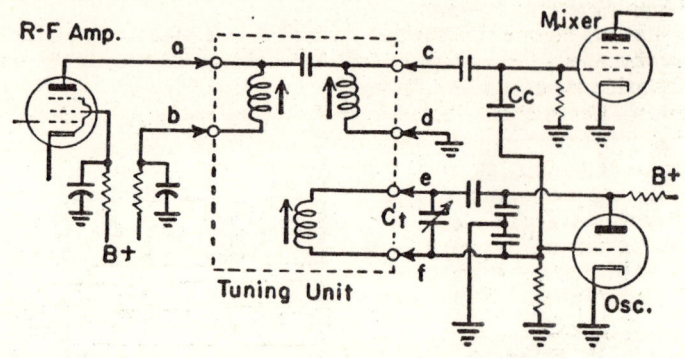

Mixer Tube

modulated wave.—*Continuous waves* of which the amplitude or the frequency is periodically varied in accordance with a signal. *See illustration.*

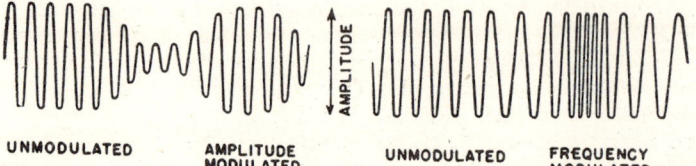

Modulated Waves

modulation.—The process of varying the *amplitude* or the *frequency* of a wave in accordance with a signal. The wave may be a radio carrier or a guided carrier current. The signal may be that corresponding to voice or music, or to key and buzzer interruptions, or to picture transmission.

modulation capability.—The greatest *percentage modulation* that may be used without appreciable distortion.

modulation distortion.—An increase in *percentage modulation* of a received signal, due to the signal working on the bend of the grid voltage-plate current characteristic of a radio-frequency amplifying tube and being affected by rectification or *detection*.

MODULATION FACTOR

modulation factor.—The variation in amplitude of a *modulated wave* from its mean value, expressed as a ratio to the mean value or as a percentage.

modulation frequency.—The frequency which is impressed upon a *carrier current* or a *carrier wave* in the process of *modulation*. The frequency which conveys the signal or intelligence to be communicated.

modulation-frequency ratio.—The ratio of the *modulation frequency* to the *carrier frequency*.

modulation meter.—A device which indicates the *percentage modulation*. Generally an *oscilloscope* showing the wave form of the modulated carrier, or a meter indicating maximum and minimum values of audio frequency wave forms.

modulation percentage.—See *percentage modulation*.

modulation transformer.—A transformer which couples a modulating tube or other modulating device to the oscillator or amplifier whose output is to be modulated.

modulator tube.—1—A vacuum tube which impresses signal variations on the output of an oscillator or amplifier system. Usually a tube with its plate resistance forming part of the power circuit of the oscillator or amplifier. Signal voltages on the grid of the modulator tube vary its effective plate resistance and there are corresponding variations in output power of the oscillator or amplifier.

molded insulation.—Insulation which is shaped under the influence of heat and pressure.

molded mica.—Small mica sheets bound together with shellac or other cementing material.

molecular theory of magnetism.—The theory that the particles forming a magnetic material are like small magnets and when acted upon by a magnetic field they align themselves with all negative poles in one direction and all positive poles in the opposite direction, thus forming negative and positive poles at the ends of the body. *See illustration*.

molecule.—The smallest particle of a substance which can exist and still have all the attributes of the substance. If the substance is a compound, further breaking up results in chemical change. If the substance is an element, the molecule is the same as an *atom*.

monaural.—Affecting only one ear of a listener. Compare *binaural*.

monel metal.—A natural alloy composed of nickel, copper and iron.

monitor.—A loud speaker, a headset or a receiver operated from the output of a transmitter or amplifier and used as an aural check on performance. Sometimes the word is used as meaning a *frequency meter*, *modulation meter*, etc., used in observing a transmitter's performance.

monochromatic.—Of only one color.

monochromatic sensitivity of photocell.—Sensitivity of a photocell to a limited range of light frequencies, or to one color.

MONOCHROME

monochrome.—A word which correctly describes pictures in a single color or tone, but which is used also to describe television pictures in black and white or without color.

Monoscope.—A tube producing a design suitable for testing the performance of a television receiver.

M. O. P. A.—An abbreviation for *master oscillator, power amplifier*.

morning glory horn.—A type of *exponential horn* commonly used with public address loud speakers. *See illustration*.

Morse code.—A system of signals used in American wire telegraphy.

mosaic.—The sensitive surface of an Iconoscope on which the picture image is focused.

mother.—A sound record disc formed from a shell which is produced by copper plating the *metal master*. Used for the production of the *stamper*.

motional impedance.—The vector difference between the normal impedance and the blocked impedance of an *electro-acoustic transducer*.

motor.—A machine which changes electrical power into mechanical power.

motorboating.—Pulses of sound at low *audio frequencies* produced from a loud speaker as a result of *feedbacks* between amplifying

Molecular Theory of Magnetism   Morning Glory Horn   Moving Coil Instrument

stages. The feedbacks generally occur through resistance couplings in power unit voltage dividers.

mouth of horn.—The larger end or outer end of a loud speaker horn.

movietone.—Sound pictures employing the *variable density system*.

moving armature loud speaker.—A *moving iron loud speaker*.

moving coil instrument.—A *direct current* type of measuring instrument consisting of a coil carrying all or a definite fraction of the current to be measured and mounted in the field of a permanent magnet. Reaction between the permanent field and the field set up by the coil causes movement of the coil and an attached pointer. *See illustration*.

MOVING COIL LOUD SPEAKER

moving coil loud speaker.—A loud speaker in which the radiating member is attached to a small coil located in the field of an electromagnet, this coil carrying the signal currents. The varying current in the coil produces around it a varying field which reacts with the field of the electromagnet to move the small coil and radiating member in unison with the signal. *See illustration.*

A permanent magnet speaker operates similarly, but has a strong permanent magnet taking the place of the electromagnet.

moving iron instrument.—A current measuring instrument in which one or more pieces of soft iron are caused to move by the magnetic field of a coil, thereby moving the indicating pointer. Used for *alternating current* measurements. Includes *plunger, vane* and *repulsion* types of instrument.

moving iron loud speaker.—A loud speaker in which the sound producing member is vibrated by movement of a piece of iron which is a part of the magnetic circuit energized by signal currents. Includes *magnetic, electromagnetic* and *balanced armature loud speakers.*

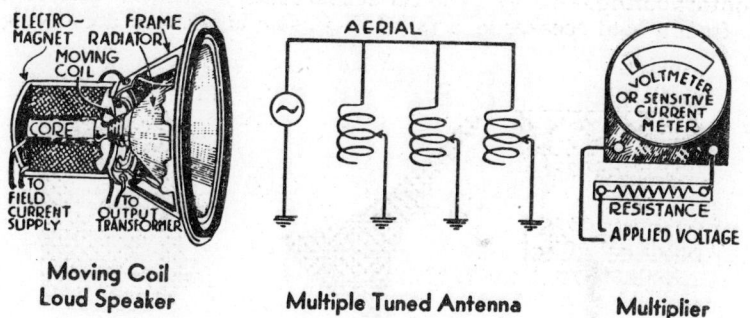

Moving Coil Loud Speaker · Multiple Tuned Antenna · Multiplier

mu (μ).—Greek letter symbol for *amplification coefficient* of a tube, or for *permeability* of a magnetic material. Also as a prefix meaning the one-millionth part of a unit.

mu-factor.—In a tube having the current of one electrode affected simultaneously by the potentials of two separate electrodes, the ratio of change between these two potentials which will maintain a constant current in the first electrode considered.

multi-layer coil.—A *honeycomb coil* or a coil having *a banked winding.* Any coil in which the winding consists of two or more superimposed layers of wire.

multiple condenser.—A *gang condenser.*

multiple connection.—A connection in a *parallel circuit.*

multiple regeneration.—*Regeneration* applied to two or more radio frequency amplifying stages, or to the detector and one or more amplifying stages.

MULTIPLE IMAGES

multiple images.—One or more relatively faint lines appearing at the right of vertical or sloping lines or edges of objects in television pictures. The cause is damped oscillation in circuit elements which carry video signals.

multiple-series.—Several units or parts connected in a number of series circuits, these circuits then being connected together in multiple or a *parallel circuit*.

multiple tuned antenna.—An antenna system in which inductances are connected at several points between the aerial and the ground or counterpoise, these inductances tuning with the antenna capacity at the operating frequency. *See illustration.*

multiplier.—A resistance placed in series with a *voltmeter* to increase the range of the meter. *See illustration.*

multipolar.—Having more than two field poles.

multipurpose transistor.—Two or more transistors assembled in a single unit but with terminals allowing their use in separate circuits.

multi-stage amplifier.—An amplifier consisting of two or more stages of *cascade amplification.*

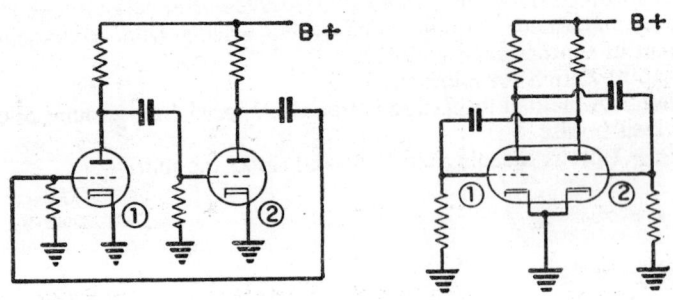

Multivibrators

multivibrator.—An oscillator employing two sets of tube elements with the output of each connected or coupled to the input of the other. Coupling may be through capacitors between plates and grids or through a common cathode resistor. Free-running frequency is determined by the time constant of a capacitor-resistor combination in either grid circuit. The multivibrator is easily synchronized by a voltage of desired frequency, near the free-running frequency, which is applied to either grid or to a cathode.

musical oscillator.—An arrangement of circuits oscillating at radio frequencies and producing various audible *beat frequencies* to form musical notes controlled in pitch by altering the circuit constants, usually by alteration of capacity in the oscillatory circuits.

mutual characteristic.—A graph showing the relation between *grid voltage* and *plate current* in a tube.

mutual conductance.—A measure of the alternating plate current which results from application of a given alternating (signal) voltage to the control grid of a vacuum tube. Mutual conductance in *mhos* is equal to the ratio of small changes of plate current to the changes of grid voltage which produce them, the plate voltage remaining constant; or is equal to the alternating plate current divided by the alternating grid (signal) voltage. Also equal to the *amplification coefficient* divided by the *plate resistance*. The value in mhos generally is reduced to *micromhos*. The symbol is G_m.

mutual inductance.—A property of coupled circuits (usually coils) by which each produces in the other an *electromotive force* whenever there is a change of current and resulting change of flux in either circuit. Measured in *henrys*. Mutual inductance exists in addition to the *self-inductance* of each separate circuit and is equal to the increase in total inductance of two circuits when they are coupled. The mutual inductance in henrys is equal to the number of volts induced in one circuit by a current changing at the rate of one ampere per second in the other circuit. The symbol is M.

mutual induction.—Production of an *electromotive force* in one circuit by movement through it of a field arising from a changing current in another nearby circuit.

mv.—Abbreviation for *millivolt*.

mycalex.—A molded insulating material prepared from ground mica and lead borate.

myria-.—A prefix meaning ten thousand times the unit.

N

N.—Symbol for number of turns in a winding.
n.—Symbol for speed, usually in revolutions per second.
napier.—A *neper*.
Napierian or Naperian logarithm.—A system of logarithms based on natural logarithms. See *logarithm*.
National Electrical Code.—A set of rules governing construction and installation of electrical devices and wiring systems; rules adopted by the National Board of Fire Underwriters, sponsored by the National Fire Protection Association, and approved by the American Standards Association.
natural frequency.—A *resonance frequency* which is determined by the effective inductance and effective capacity in a circuit. In an antenna circuit it is the lowest resonance frequency with no added lumped inductance or capacity.
natural impedance.—*Iterative impedance.*
natural logarithm.—See *logarithm*.
natural period.—The *period* corresponding to the *natural frequency*.
natural wavelength.—The *wavelength* corresponding to the *natural frequency*.
navy socket.—A vacuum tube socket with a high cylindrical shell slotted at one point for the locating pin on a tube base. See *illustration*.
N.E.C.—Abbreviation for National Electrical Code.
neck shadow.—A darkened area on television pictures caused by the electron beam striking inside the neck of the picture tube before reaching the viewing screen.
needle gap.—Two pieces of sharply pointed metal used for measurement of high voltages by breakdown of the air resistance and passage of a spark between them.
needle scratch.—Audio frequency voltages produced by movement of a phonograph pickup needle over unintentional irregularities on the surface of the record.
negative.—Descriptive of any point in an electric circuit at which are more free electrons than at other (relatively positive) points in the same circuit, and away from which electrons tend to flow toward places which are positive. Also any body which is negatively charged due to an excess of electrons.
negative bias.—A constant direct voltage applied between control grid and cathode of a tube, in polarity which makes the grid negative with respect to the cathode. The usual purpose is to pre-

NEGATIVE CARRIER

vent maximum positive amplitudes of applied signals from making the grid positive with respect to the cathode. This requires bias voltage at least equal to signal amplitude.

negative carrier.—A *negative ion.*

negative charge.—The quantity of electricity in a body which is at a negative potential with reference to another body which carries an equal positive *charge.*

negative compliance.—The condition under which an initial small movement of a mechanical part results in still further movement in the same direction. A decreasing amount of stiffness.

negative corpuscle.—An *electron.*

negative electrification.—The condition of a body which is carrying an excess of negative electrons. See *electrification.*

negative glow.—A luminous discharge existing near the cathode of a gaseous tube in which there is a flow of current accompanied by *ionization.*

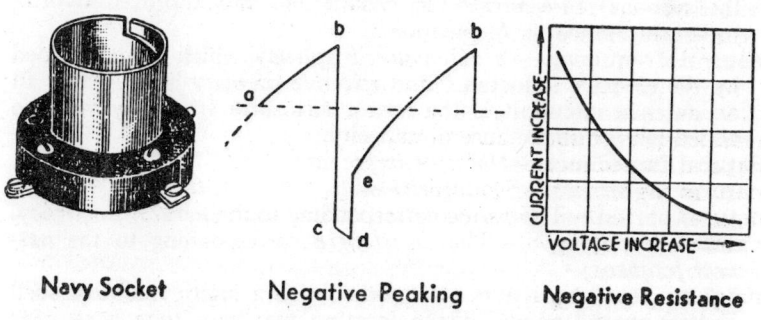

Navy Socket   Negative Peaking   Negative Resistance

negative grid.—The condition in which a control grid is provided with *negative bias.*

negative image.—An image or picture in which lights and shadows are reversed in position from those of the original.

negative ion.—An atom having an extra electron, having a negative charge. An ion which moves toward an *anode.* An anion.

negative peaking.—Addition of a brief pulse in negative polarity to the sudden change of voltage in a sawtooth wave. Assists in forming sawtooth currents in some kinds of connected inductive circuits, or in altering a drive voltage.

negative picture.—A television picture on which areas that should be dark are light and those which should be light are dark. The appearance is that of a photographic negative.

negative pole.—1—The end of a magnetic needle which points toward the south; the south pole. The end of a magnet into which pass the *magnetic lines of force* from the external field.

negative resistance.—1—A condition in an electric circuit wherein an increase of voltage is accompanied by a decrease of current.

NEGATIVE TEMPERATURE COEFFICIENT

A characteristic of an electric *arc* and of a *dynatron* tube. A condition resulting from *secondary emission* in a tube. *See illustration.* 2—The condition under which there is a *feedback* of sufficient energy from a tube's plate circuit to completely overcome the effect of resistance in the grid circuit and to allow sustained oscillations.

negative temperature coefficient.—A characteristic of a substance in which the resistance becomes less as the temperature increases. Carbon and most liquid conductors have negative temperature coefficients.

negative transmission.—Television carrier modulation with which tips of all sync pulses are at maximum amplitudes, while portions representing brightest areas of pictures are at minimum amplitudes. The standard method of transmission in the United States and some other countries.

negative wire.—A wire connected to the *negative* side of the source.

negatron.—A four-element tube having two plates and exhibiting the property of *negative resistance*.

neon arc.—A light source in which ionization allows a *glow discharge* through neon gas. The intensity of the glow is increased by concentrating the effect within a small area.

neon lamp or bulb.—A glass bulb containing two electrodes in neon gas at low pressure. With voltage in excess of a critical value across the electrodes, ionization takes place and the cathode (connected to the negative side of the voltage source) becomes covered with a pink glow.

neon oscillator.—A *neon lamp* and a condenser incorporated in a circuit of which the time constant may be varied over a wide range by adjustment of the condenser's capacity. Variation in the frequency of current fluctuations results from change of condenser capacity. *See illustration, page following.*

neper.—A *transmission unit* in the napierian system. The unit for power measurements is one-half the natural logarithm of the ratio between the two powers considered.

net reactance.—The difference between the *inductive reactance* and the *capacitive reactance* in an alternating current circuit. The difference is positive when the inductive reactance is the greater and is negative when the capacitive reactance is the greater of the two.

neutral.—Neither negative or positive. A point which is positive with reference to some points and negative with reference to others in an electric or a magnetic circuit.

neutral density face plate.—A television picture tube face plate in which about one-third of light reaching one side is absorbed in passing through the glass. Light passing outward from the viewing screen suffers one such absorption, but external light reflected back from the screen is subjected to two absorptions and appears less bright.

neutral relay.—A relay which operates upon change of current without regard to polarity.

neutral wire.—The center wire of a *three-wire system*.

neutralizing.—*Balancing.*

neutrodyne receiver.—A receiver using *tuned radio frequency amplification* in which oscillation in the radio frequency stages is prevented by balancing the tubes' internal capacities with external capacities, allowing a *feedback* equal in effect but opposite in phase to the feedback through the tubes. The tendency toward oscillation which would result from tube feedback is balanced by an opposite effect from the external feedback. *See illustration.*

nichrome.—An alloy of nickel, chromium and iron having high electrical resistance and the ability to withstand high temperatures.

nickel iron.—An alloy of nickel and iron having high permeability, used for *cores* in transformers and chokes.

nickel silver.—A *resistance* alloy composed of copper, nickel and zinc. German silver.

Nicol prism.—A device for producing a beam of *plane polarized light*. Two pieces of transparent calcite, formed by cutting diagonally through a natural crystal, are cemented together. A light beam entering the prism is subjected to *double refraction*, one beam of plane polarized light passing on through the prism while

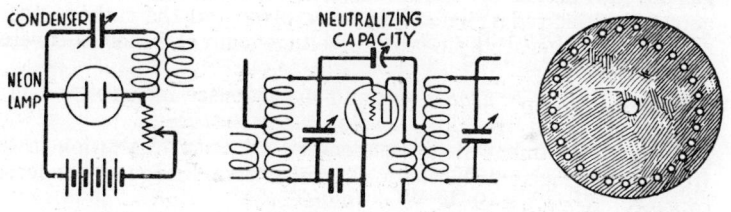

Neon Oscillator   **Neutrodyne Receiver**   **Nipkow Disc**

the second beam (of ordinary light) is reflected from the cemented joint and is deflected to one side.

Nipkow disc.—A flat, round plate with a series of openings spaced at equal angles near the outer edge, successive openings being at gradually increasing distances from the center so that the series forms a spiral. At a *television* receiver, rotation of the disc allows viewing or scanning a television lamp to form the small elementary areas of an image, while at a transmitter the disc either directs a light beam over all the elementary areas making up an image or directs illumination from an image to a photocell. A scanning disc. *See illustration.*

nodal point.—A *node*.

node.—A point in a series of sound waves, radio waves or other vibrations at which there is no motion of the transmitting medium in any direction. A point at which *direct* and *reflected waves* neutralize each other, resulting in no motion of the transmitting medium. A point on a conductor, such as an antenna, at which there is zero voltage or zero current. *See illustration, page following.*

Nodon or Noden valve.—An *electrolytic rectifier*.

noise.—Any sounds resulting from irregular or confused frequencies of wave motion. Television picture noise refers to picture distortion resulting from any voltages which, in sound receiver circuits, would cause audible noise. Transistor noise voltage results from shifting of electrons and holes in the transistor crystal. See also thermal noise and tube noise.

noise figure.—The ratio of noise signal power actually existing, to the noise power which would exist were there no tube, thermal, or other noises produced in receiver circuits. A measure of internal receiver noise as distinct from noise entering through the antenna.

non-conductor.—Any material used to obstruct flow of electricity; an *insulator*.

non-inductive circuit.—A circuit in which the value of *inductance* is negligible in comparison with the *resistance*.

non-inductive resistor.—A resistance unit having negligible inductance. A resistor having a *non-inductive winding;* or one made from a straight conductor, a mass of carbon, etc.

non-inductive winding.—A winding in which half the turns run in one direction and the other half in the opposite direction around the core. The field of one half then neutralizes the field of the other half and there is very little *self-inductance*.

non-linear.—A relation between electrical quantities such that change in one quantity is not exactly proportional to change in another controlling quantity through the working range of a device.

non-magnetic.—Descriptive of any substance which does not become a *magnet;* including practically all common materials except iron and steel.

non-oscillatory circuit.—An *aperiodic circuit;* a circuit which does not exhibit resonance.

non-reactive load.—A load in which the current and voltage are *in phase*.

normal.—A line or a plane which is perpendicular to another line or plane, or to a tangent of a curve.

normal atom.—An *atom* in which the positive and negative charges are equally balanced.

normal cut.—An *X-cut* for a quartz crystal.

normal emission.—*Photoelectric emission* which changes gradually and only slightly as the light frequency is varied over the visible spectrum, as distinct from *selective emission. See illustration.*

normal impedance.—The impedance at the terminals of the electrical or the mechanical system of an *electro-acoustic transducer* when the other system is connected to its normal load.

normal input voltage.—The r-m-s voltage of a signal which is modulated at 400 cycles, *percentage modulation* of 30, applied to a receiver's input during standard tests.

normal radio field intensity.—The *radio field intensity* of a carrier wave modulated at 400 cycles with *percentage modulation* of 30, which allows the *normal test output* from a receiver.

normal test output.—A power of 0.05 watt developed in a *noninductive resistor* carrying only alternating current of *audio frequency* and connected across the loud speaker terminals of a radio receiver. The resistance is selected of a value which allows

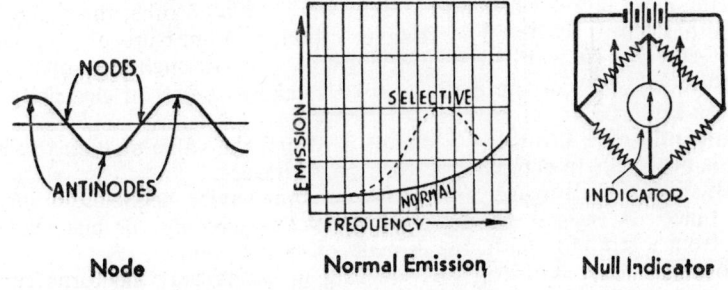

Node　　　**Normal Emission**　　　**Null Indicator**

maximum power output from the receiver.

north pole.—The end of a magnetic needle which points north.

notation by powers of 10.—See *standard notation*.

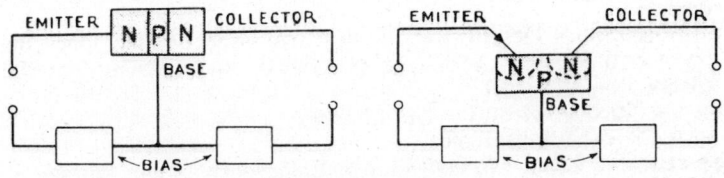

N-P-N Transistor

no-voltage release.—A device including a magnet which holds a switch in its operating position until the line voltage drops to zero, whereupon a spring opens the switch.

***N*-phase.**—More than one phase; *polyphase*.

n-p-n transistor.—A junction transistor in which emitter and collector are of n-type with between them a base of p-type.

NTC resistor.—A resistor having large temperature coefficient of resistance, causing resistance to decrease rapidly with rise of temperature due to current flow.

N.T.S.C.—Abbreviation for National Television System Committee.

n-type crystal.—A transistor crystal to which has been added an impurity whose atoms have more valence electrons than atoms of a pure crystal, thus providing negative electrons which may move in the transistor crystal.

null indicator.—Any device which indicates zero current flow or an equality of potentials, as required in bridge measurements. *Headphones, galvanometers, vacuum tube voltmeters, etc.*, may be used as null indicators. *See illustration.*

null (test or measurement) method.—An electrical test in which the pointer of an instrument is brought to zero when a balance or other desired condition is attained.

O

octal base.—An electron tube base having spaces for eight contact pins, with pins at all or at only some of the possible positions.

oersted.—The C.G.S. unit of *reluctance*. The opposition offered to flow of magnetic lines of force by one cubic centimeter of a vacuum. The C.G.S. oersted is equal to 0.796 oersteds in the English system of units.

ohm.—The practical unit of electrical *resistance* to flow of current. The resistance which allows one *volt* potential difference to cause a current flow of one *ampere*. The approximate resistance of 1,000 feet of number 10 copper wire or 380 feet of number 14 copper wire. The *international ohm*.

ohmic drop.—The *potential difference*.

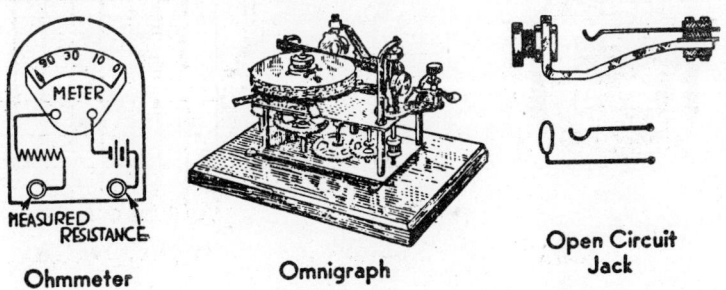

Ohmmeter **Omnigraph** **Open Circuit Jack**

ohmic resistance.—The resistance to flow of *direct current*. That part of the total opposition to flow of any current that is caused by the conductor's material, length, cross section and temperature without consideration of *inductance* or *capacity*.

ohmmeter.—An instrument for measuring and indicating the value of a *resistance* directly in ohms. In one type the unknown resistance is proportional to the deflection of a meter in a circuit including a battery *(see illustration)* while other types are based on *slide wire bridges*, etc.

Ohm's law.—The law which defines the relation between *electromotive force* in volts, *current* in amperes and *resistance* in ohms of a circuit carrying a steady direct current. Using the symbols: E for volts, I for amperes and R for ohms, the law is expressed as follows: $E = I \times R \quad I = E/R \quad R = E/I$

ohms per volt.—The quotient of dividing total internal resistance of a voltmeter, plus any multiplier, by the number of volts at full scale. The greater the number of ohms per volt the less current is taken from measured circuits by the meter and the greater is meter sensitivity.

oiled cloth.—Linen or cotton coated with linseed oil.

omega (ω or Ω).—Greek letter symbol for *resistance* in ohms, also for *angular velocity*.

omnigraph.—An instrument for producing audible *code* signals from a buzzer operated with current suitably interrupted by a perforated tape or disc of insulating material. *See illustration, page preceding.*

ondometer.—A *frequency meter*.

one hundred per cent modulation.—*Complete modulation.*

opaque.—Not permitting passage of light, not transparent.

open antenna.—An antenna consisting of an elevated aerial and the ground underneath. A *capacity antenna*.

open circuit.—A circuit which is not electrically continuous and in which current can not flow.

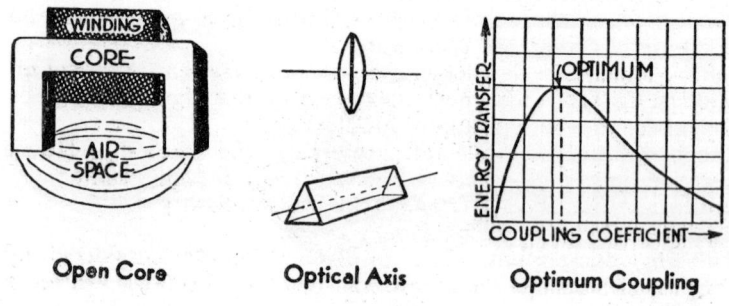

Open Core   Optical Axis   Optimum Coupling

open circuit characteristic.—A graph showing relations between voltage, current or other electrical values when external circuits are disconnected or open.

open circuit current.—Magnetizing current.

open circuit jack.—A *jack* which normally opens a circuit and closes it only upon insertion of the *plug*. *See illustration, page preceding.*

open circuit voltage.—The voltage across the terminals of a source when no current is flowing from it.

open core.—A magnetic *core* extending little if any beyond the ends of its winding, the *magnetic circuit* containing a considerable amount of air or non-magnetic material. *See illustration.*

open fuse.—A fuse wire which is not covered. A *link fuse*.

open line.—A resonant line whose conductors are not conductively connected together at the end farthest from the voltage source.

open wire transmission line.—A transmission line whose conductors are not embedded in or enclosed by continuous insulation, but are spaced apart by insulators at regular intervals along the length of line.

operation.—See *class A, B, C operation*.

opposite phase.—A *phase difference* of 180 electrical degrees.

optical axis.—In a lens or a crystal, the imaginary straight line following the path taken by the light rays which are not changed in direction in passing through the lens or crystal. The path followed by rays which are not *refracted* or *reflected*. The *Z-axis* in a piezo-electric crystal. A straight line through the center of curvature of a lens. *See illustration*.

optical center.—A point within or near a lens and on the *optical axis*, where rays of light are not changed in direction.

optics.—The science which deals with light and with vision.

optimum coupling.—The degree of coupling or the *coupling coefficient* with which there is maximum transfer of energy. *See illustration*.

optimum reverberation.—The degree of *reverberation* which results in maximum apparent loudness of a sound without greatly lessening the *intelligibility*.

ordinate.—A distance measured vertically on a graph to locate points on a curve. See *coordinates*.

orient.—To position the pickup elements and other conductors of an antenna to allow maximum pickup of carrier signal energy, or sometimes to reduce pickup of interference.

orthochromatic.—Capable of preserving the natural relations between lights and shades or between colors in photography.

oscillating.—See also under *oscillation* and *oscillatory*.

oscillating circuit.—An *oscillatory circuit*.

oscillating component.—The *oscillatory current*, considered by itself, which exists simultaneously with a direct current in a circuit.

oscillating crystal.—A *piezo-electric crystal*.

oscillating current.—An *oscillatory current*.

oscillating tube.—A tube in whose plate or anode circuit, and in whose grid or control circuit there is *sustained oscillation*. A tube which is carrying *oscillatory current*.

oscillation.—1—A high frequency alternating current, especially such a current existing in an *oscillatory circuit* when the energy swings back and forth between inductance and capacity. 2—The condition in which there is *sustained oscillation* in a circuit or an amplifier. 3—See definitions under *oscillating* and *oscillatory*.

oscillation constant.—The square root of the product of an *inductance* and a *capacity* which together cause *resonance* at the frequency considered.

oscillation frequency.—The frequency at which a circuit exhibits *resonance*.

OSCILLATOR

oscillator.—A device without rotating parts which produces alternating voltages and currents of a frequency depending on the electrical characteristics of the device. An *audio frequency oscillator* or a *radio frequency oscillator*.

oscillatory.—See also under *oscillating* and *oscillation*.

oscillatory circuit.—A low resistance circuit in which periodically reversing currents will flow between the inductance and the capacity at the *natural frequency* of the circuit when an e.m.f. is applied. A circuit in which the square of the resistance in ohms times the capacity in farads is less than four times the inductance in henrys. *See illustration.*

oscillatory current.—A high frequency current flowing back and forth between the inductance and the capacity in an *oscillatory circuit*. Usually such a current of steadily diminishing amplitude.

oscillion.—A *three-element tube*.

oscillogram.—A record of readings from an *oscillograph*. *See illustration.*

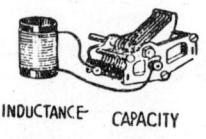

Oscillatory Circuit

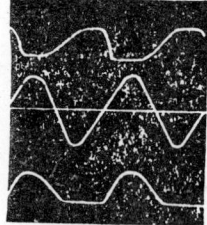

Oscillogram

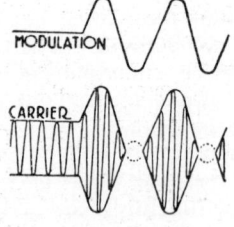

Overmodulation

oscillograph.—An oscilloscope.

oscilloscope.—An instrument containing a cathode-ray tube together with a horizontal sweep oscillator, suitable amplifiers for horizontal and vertical sweep voltages, various operating controls, and usually a self-contained power supply. Designed to allow display on the viewing screen of the cathode-ray tube of waveforms and other changes of voltage with respect to time when the voltages are applied to input terminals of the instrument.

outdoor antenna.—An antenna system having the *aerial* conductor erected out of doors.

output.—The power, current, voltage or field delivered by a source into external circuits or to any form of load.

output choke.—An *inductance coil* in the plate circuit of the last tube or tubes in an *audio frequency amplifier*, arranged so that voltage changes across the coil are carried through a coupling condenser to a loud speaker. A means for preventing direct current in the plate circuit from flowing through loud speaker windings.

OUTPUT CIRCUIT

output circuit.—A circuit through which energy from a source is delivered to external circuits or to the load.

output impedance.—*Internal output impedance.*

output meter.—A meter especially suited for measurement of audio-frequency voltages. Often a rectifier meter.

output resistance.—1—The *plate resistance* of a vacuum tube. 2—*Internal resistance.*

output transformer.—A transformer which couples the *output* of a source, such as a tube's plate circuit, to the *load* on the source.

overall response.—A frequency response affected by combined gains and attenuations of several circuits, elements, or amplifier stages.

overall selectivity.—The net effective *selectivity* of an entire receiver or amplifier; the result of the cascaded selectivities of the separate circuits or stages. Equal to the nth root of the product of n separate selectivities.

overcoupled transformer.—A tuned high-frequency transformer in which coupling between secondary and primary is so close as to cause resonance at two frequencies whose separation increases with coupling. A frequency response has two peaks. Compare double hump resonance.

overload level.—The input power or output power at which a device suffers in performance because of distortion, breakage, overheating, burnout, etc.

overload relay.—A relay which opens a circuit upon a flow of abnormally high current.

overloading.—Operation of any amplifying tube or detector tube at control grid signal voltages having peak values greater than the negative *grid bias,* the result being that the *grid voltage* becomes positive on peak signal amplitudes with consequent *harmonic distortion*.

overmodulation.—Modulation in which *modulation frequency* amplitudes are periodically made greater than the *carrier current* amplitude, resulting in distortion due to complete stoppage of radio frequency oscillations during such periods. *See illustration, page preceding.*

override.—To connect a fixed direct voltage, usually from a battery, between the two sides of a circuit for automatic gain control or volume control. Grid voltages and tube gains thus are held nearly constant during service operations.

overtone.—A *harmonic* of a sound frequency.

oxide coated filament.—A tube filament consisting of a platinum alloy covered with oxides of barium, calcium and strontium. Has large *electron emission* at low temperatures.

oxide rectifier.—A *copper oxide rectifier.*

P

P.—Symbol for *average* power in watts.

p.—Symbol for *instantaneous* value of power in watts.

packing of microphone.—Reduction in resistance and response of a *carbon microphone* as a result of excessive pressure on the contacting particles.

pad.—Resistors in series or parallel combinations used between instruments, receivers, or other devices to provide impedance matching or to uniformly attenuate a range of frequencies.

padder capacitor.—A capacitor, usually adjustable, in series between inductance and capacitance of a resonant circuit to allow varying the circuit capacitance for tracking during tuning of the circuit.

pairing.—Drawing together or overlapping of horizontal lines in successive fields of television pictures, due to poor vertical synchronization.

pancake coil.—A *spiderweb coil*. Any coil formed by a spirally wound conductor.

panchromatic.—Capable of preserving or reproducing the natural contrasts between colors which are visible to the human eye.

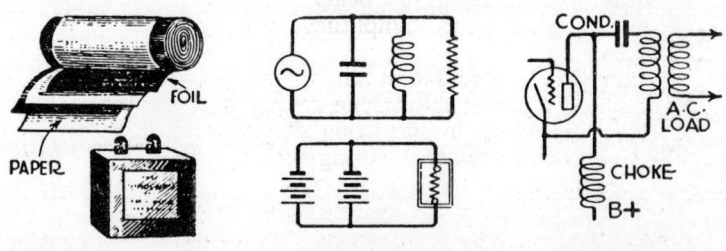

Paper Condenser   Parallel Circuit   Parallel Feed

paper capacitor.—A capacitor having plates of thin metal foil separated by dielectric of one or more layers of waxed or otherwise treated paper, the assembly being rolled into a cylinder and sometimes flattened, then encased in cardboard, plastic, or metal for protection.

parabolic current or voltage.—In convergence circuits of color television receivers, currents or voltages whose amplitude is varied proportionately to horizontal or vertical deflection of electron

PARAFFIN

beams, for the purpose of maintaining (dynamic) convergence toward sides, top and bottom of the viewing screen.

paraffin.—A vegetable wax used for its insulating and dielectric properties. Has low radio frequency losses. Dielectric constant 2.0 to 2.5.

parallel circuit.—A circuit including several parts, either sources or loads, with one end of every part connected to one of a pair of conductors and the opposite end of every part connected to the second conductor, the pair of conductors thus having between them all the several parts so that current divides between the parts proportionately to their e.m.fs. or to their impedances. A shunt circuit or a multiple circuit. *See illustration.*

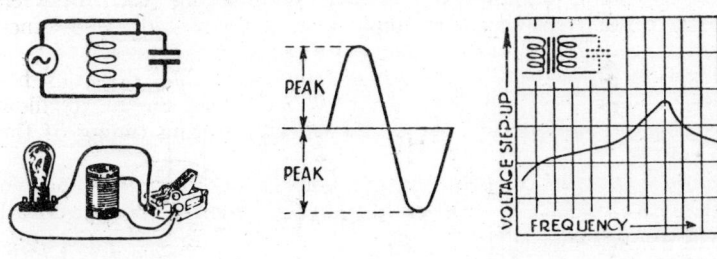

Parallel Resonance   Peak Value   Peaked Transformer

parallel cut.—A *Y-cut* for a quartz crystal.

parallel feed.—A connection of the direct current power supply for the plate of a tube to the plate through a choke coil, the audio frequency or radio frequency alternating plate current then taking a separate path through a condenser to a load winding or coupler. Compare *series feed*. *See illustration, page preceding.*

parallel feed oscillator.—An *oscillator* circuit employing *parallel feed*.

parallel resistance.—1—A resistance in a *parallel circuit*. 2—An *equivalent resistance* (in parallel).

parallel resonance.—The condition in a circuit containing capacity and inductance in parallel with each other, and of such values that they produce *resonance* at the frequency existing in the main circuit. The *impedance* offered to flow of current in the main circuit is very great. The total current circulating in the capacity and inductance is greater than the current in the main circuit, and is in phase with the applied e.m.f. Compare *series resonance*. *See illustration.*

paramagnetic.—Having the ability to become a magnet. Having a *permeability* greater than that of air or greater than unity. Iron and steel are the chief paramagnetic materials. See *diamagnetic.*

parasites.—*Static* disturbances.

PARASITIC ELEMENT

parasitic element.—Any reflector or director used with a dipole antenna.

parasitic oscillation.—High-frequency oscillation in inductances and capacitances not intended to be resonant. Such oscillation occurs in stray and distributed inductances and capacitances and in tube inductances and capacitances.

partial tones.—The separate frequencies, or *pure tones*, which enter into a *complex wave* or tone. The fundamental and the separate harmonics each are partial tones.

Paschen's law.—A law stating the relation between the *ionization potential* in a gaseous tube and the product of gas pressure and distance between electrodes.

pass band.—A range of frequencies throughout which there is desired or acceptable amplification or gain, or in which gain is not less than 0.707 times the maximum.

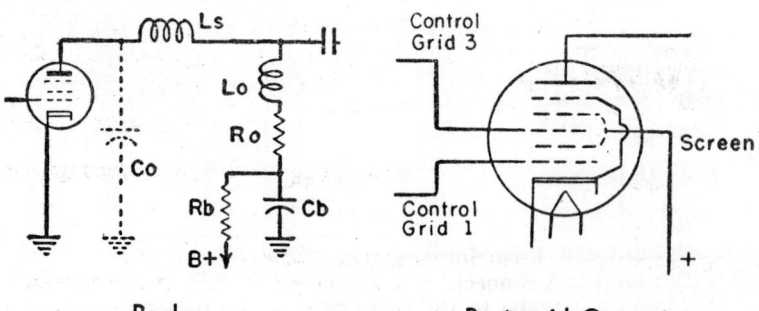

Peakers   Pentagrid Converter

passive transducer.—A *transducer* supplying no power from itself but taking from the first system all of the power supplied to the second system connected to the transducer.

pattern generator.—A signal generator for producing on the viewing screen of a television picture tube various lines or dots in geometric patterns useful during service operations.

peak inverse voltage.—The peak voltage applied to a device, such as a rectifier, in a direction opposite to that of normal current flow.

peak separation.—The difference in frequency between the two resonant peaks of the two coupled circuits as used in a *band selector circuit*. See *double hump resonance*.

peak value.—The *amplitude* of an alternating quantity. The greatest value of current, voltage, power, etc., reached during an *alternation*. With sine wave currents the peak value is equal to 1.414 times the effective value, and the effective value is equal to 0.707 times the peak value. The symbol for peak voltage is E_m, for peak current is I_m and for peak power is P_m. See *illustration*.

peak voltmeter.—A voltmeter which indicates the maximum value reached by an alternating voltage during an alternation. A *vacuum tube voltmeter*.

peaked transformer.—An *audio frequency transformer* in which the combination of inductance and distributed capacity produces *resonance* and high sensitivity at some one frequency.

peaker.—A small inductor, sometimes adjustable, which resonates with capacitances in video detector or amplifier output circuits to provide increased load and gain in the frequency range required for high-frequency compensation.

peaking control.—A television receiver adjustment for varying the degree of negative peaking on a sawtooth voltage fed to the horizontal output amplifier.

peak-to-peak value.—A potential difference or current change measured from maximum positive to maximum negative amplitudes of a wave. The sum of positive and negative peak amplitudes.

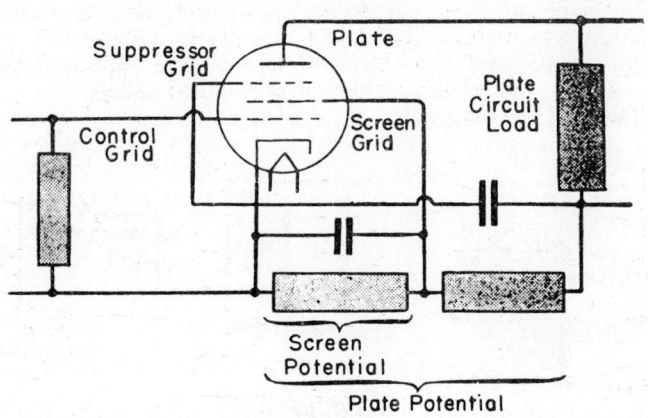

Pentode

pedestal.—In the composite television signal, black level voltages preceding and following each horizontal sync pulse during the horizontal blanking period.

Peltier effect.—The change of temperature caused at a joint between two metals by flow of current through the joint.

pentagrid converter.—A single tube which, in a superheterodyne receiver, combines the functions of *oscillator* and *first detector* by employing *electron coupling* to combine an oscillator frequency with the frequency or amplitude of a signal. A grid and anode nearest the cathode produce the oscillator frequency in the single electron stream which is modulated by the signal in other elements between cathode and plate.

pentode.—A tube in which are five elements consisting of the cathode, the first or control grid, the second or screen grid, the third or suppressor grid, and the plate. The suppressor grid, maintained at or close to cathode potential by internal or external connection, is highly negative with respect to the plate and repels back to the plate most of the secondary emission electrons that leave the plate and otherwise would pass to and into the positive screen grid. Most pentodes are designed for voltage amplification, although some are power amplifiers.

percentage modulation.—The percentage of its mean value by which a *modulated wave* varies from that mean. The ratio (expressed as a percentage) of half the difference between maximum and minimum *amplitudes* to the average amplitude of a modulated wave. *See illustration.*

period.—1—The time in seconds required to complete one *cycle* of an alternating quantity. 2—The time required for the pointer of an instrument to make two full swings, one in each direction.

periodic.—Repeating regularly in form and time.

periodic current.—An *oscillatory current* having a frequency determined by the inductance and capacity in its circuit.

periodic wave.—Any form of *wave motion* which repeats regularly, successive cycles exhibiting similar changes of values.

periodicity.—*Frequency.*

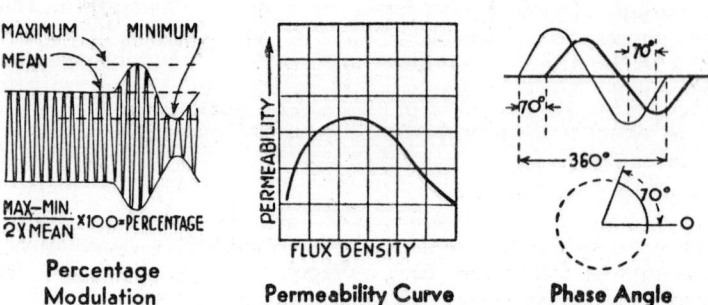

Percentage Modulation

Permeability Curve

Phase Angle

permalloy.—An alloy consisting of iron and nickel, having very great *permeability* and used for the cores of audio transformers, chokes and similar parts.

permanent magnet.—A piece of hardened steel which has been magnetized and which retains its magnetic strength.

permanent-magnet moving-coil instrument.—An instrument having a coil moving in the field of a *permanent magnet.*

permanent magnet speaker.—A speaker whose magnetic field, wherein the voice coil moves, is in a circular gap between poles of a strong permanent magnet. Compare moving coil loud speaker.

permeability.—A measure of the ability of a material to carry or conduct *magnetic flux* or lines of force. The ratio of the *flux* in lines (B) to the *magnetizing force* in gilberts or in ampere-turns per unit length of magnetic path *(H)*. The permeability of air or non-magnetic materials is 1, or unity, and the permeability of magnetic substances is measured in multiples of the permeability of air. The symbol is the Greek letter mu (μ).

permeability curve.—A graph showing the relation between *magnetic flux density (B)* and *permeability* (μ) of a substance.

permeability tuning.—Varying the inductance and resonant frequency of a circuit by means of an adjustable powdered iron core inside the coiled inductor.

permeameter.—An instrument for measuring the *permeability* of a magnetic material by testing its effect in a magnetic circuit.

permeance.—The property of a material which allows passage through it of magnetic flux. The reciprocal of *reluctance* in a magnetic circuit.

permittance.—The ability of a *dielectric* to transmit forces accompanying an electrostatic charge. The *electrostatic capacity*.

permittivity.—The *permittance* of a centimeter cube of a substance, measured between opposite faces. Equal to the *dielectric constant*.

perpendicular cut.—An *X-cut* for a quartz crystal.

persistence of phosphor.—Ability of a phosphor to emit light at a diminishing rate after the exciting electron beam has left the phosphor area.

persistence of vision.—A characteristic of the human eye in which the impression of an image remains for one-twentieth to one-tenth of a second after the real image or object is no longer being viewed.

perveance.—A tube characteristic related to ability for conducting high-frequency currents with little loss of signal strength or amplitude. With other factors constant, perveance increases with cathode area and with less spacing between cathode and the first element outside the cathode.

p.f.—Abbreviation for *power factor*.

phanotron.—A gas-filled hot-cathode rectifier tube handling large currents at moderately high voltages.

phantom antenna.—An *artificial antenna*.

phase.—The portion of a cycle through which an alternating voltage or current has progressed from whatever instant in the cycle is considered as the starting point. It is convenient to think of the voltage or current as being represented by its peak amplitude in either polarity, and to think of each complete cycle as represented by a circle around which travels the voltage or current. As with any circle, the cycle may be divided into 360 degrees, with any point at which action is presumed to commence designated as zero degrees. Then the portion of a cycle which has been completed by the voltage or current at any instant of time may be specified as so many electrical degrees with reference to the starting point or instant. If two voltages, two currents, or a voltage and a current reach the same point in their cycles at the same instant they are in phase. Otherwise they are out of phase, and the phase difference

PHASE ANGLE

between them is measured in electrical degrees. Phase, in this respect, is a time relation between alternating voltages or currents.

phase angle.—A time difference in alternating quantities, the difference being expressed in *electrical degrees*. A difference in time between similar values of two alternating quantities, or a difference between two time instants of one such quantity. *See illustration.*

phase angle difference.—The difference between ninety *electrical degrees* and the actual number of degrees by which the current in a condenser leads the applied voltage. In an ideal condenser the current would lead the voltage by ninety degrees and the phase angle difference would be zero. The phase angle difference is a measure of power loss in a condenser and it depends largely on the kind of *dielectric*. The symbol is the Greek letter theta (θ).

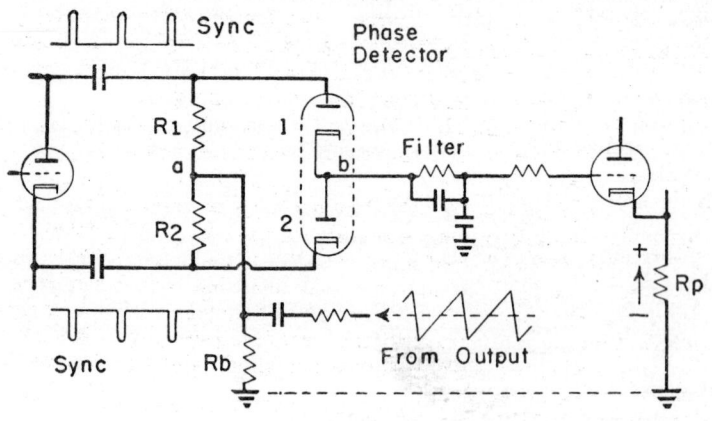

Phase Detector

phase control.—A hue control for a color television receiver.

phase detector.—A type of automatic frequency control which provides for the oscillator grid circuit a potential of one polarity when oscillator frequency increases with reference to a synchronizing frequency and of opposite polarity when it decreases, thereby correcting the error. With a twin diode phase detector the cathode of one diode and plate of the other receive two synchronizing voltages of opposite polarity, usually obtained from a phase splitter. To the remaining plate and cathode, tied together, is applied a voltage from the oscillator output or beyond. If oscillator and synchronizing voltages are in phase, their waves combine to cause equal conductions in both diodes. Otherwise the combined amplitudes are greater on one diode than the other and cause unequal conductions. Any difference of voltages across two resistors carrying the diode currents is applied as correction voltage to the oscillator. With a triode phase detector the synchronizing voltages are applied to cathode and grid, while an oscillator output voltage comes to the plate.

phase difference.—The number of *electrical degrees* separating similar values (zero or maximum) of two alternating quantities existing together and having the same *frequency*. The position of either quantity may be taken as the reference point. See *angle of lag and of lead*.

phase displacement.—An advancing or retarding of the *phase* of the current in an alternating circuit with reference to the applied voltage.

phase distortion.—A change in *phase* relation between voltage and current at certain frequencies during their passage through a transmission line or a *transducer* of any form.

phase inverter.—An inverter tube.

phase modulation.—Momentarily varying the phase of a carrier with respect to its average phase by means of a modulating voltage representing the signal to be transmitted. The action is somewhat similar to frequency modulation.

phase opposition.—A *phase difference* of 180 degrees, or one-half cycle.

phase shift.—A change of phase between input and output of an amplifier or other device, with phase of output voltage or current either lagging or leading that at the input.

phase splitter.—Usually a triode with input to the grid and outputs from both plate and cathode. Phase of plate output voltage is opposite to that of grid voltage, while cathode output is in phase with grid input voltage.

phased.—Synchronized, both in frequency and phase.

phasing links.—Conductors of such lengths and methods of connection between antenna bays as will bring all received signals into phase at the transmission line takeoff.

phenol fibre.—Fibre impregnated with a *phenolic compound*.

phenolic compound.—The basic substance of many insulating materials such as bakelite, formica, celeron, etc. It consists of phenol and formaldehyde treated with heat and pressure and moulded into various forms. Dielectric constant 4.5 to 7.5. These compounds are resistant to heat, gases, acids and moisture, while having good mechanical strength.

phi (ϕ).—Greek letter symbol for *magnetic flux*.

phone.—A contraction of "telephone" or "radio telephony".

phonetic.—Pertaining to speech sounds, to the voice and its use.

phonetics.—The science of speech; analysis and classification of speech sounds.

phonograph amplifier.—An *audio frequency amplifier* especially designed to operate with a *phonograph pickup* as a signal source.

phonograph pickup.—A device which produces *audio frequency* currents or voltages from movement of a needle over a phonograph record.

phonograph recorder.—A device for cutting into the surface of a phonograph disc the variations in form or contour representing *audio frequency* changes in a current controlling the operation.

phosphor.—A mixture of materials which emit light when struck by electrons. Used in viewing screens of television picture tubes and cathode-ray tubes. Color of emitted light depends on the kind of phosphor material.

phosphor dot.—The smallest area of phosphor emitting one color on the viewing screen of a three-gun color television picture tube. Dots of the three primary colors, blue, green and red, are arranged in triangular groups.

phosphor strip.—A very narrow band of phosphor material for one primary color, extending across the phosphor plate of a single-gun color television picture tube.

phosphorescence.—An emission of light resulting from exposure to light rays or other radiant energy and continuing after the exciting radiation has been removed. A form of *luminescence*.

phot.—A unit of *illumination* equal to one *lumen* per square centimeter.

photocell.—A device in which changes of light cause either a change of resistance, the photo-conductive effect, or else a change of e.m.f. produced in the cell, the photo-voltaic effect. To be distinguished from a phototube in which action is photoemissive.

photocell conductance.—The ratio of changes of current in a photocell to the changes of voltage which produce them. It represents the slope of the cell's current-voltage curve.

photocell current.—The entire current passing through a photocell between anode and cathode. The *primary photoelectric current* plus the current due to *gas amplification*.

photocell resistance.—The reciprocal of *photocell conductance*.

photocell sensitivity.—The ratio of changes of current in a photocell to the changes of *luminous flux* or of *radiant flux* which cause them. It represents the slope of the cell's current-illumination curve. The symbol is S. Static sensitivity is the ratio of direct current to steady light or steady radiant flux. Dynamic sensitivity is the ratio of alternating current to a varying flux. See *color sensitivity*.

photo-chemical.—Pertaining to the chemical effects of light. See *actinic ray*.

photo-conductive.—Exhibiting a change of electrical conductivity or *ohmic resistance* when acted upon by light. A quality of the element *selenium*.

photo-current.—The electric current which corresponds to an electron flow caused by the action of light on a cathode surface in a *photocell*.

photodynamic.—Relating to *radiant energy* of light.

photoelectric.—Pertaining to the effects of *light* on electric circuits.

photoelectric effect.—The emission of electrons under the influence of *light* or other *radiant energy*.

photoelectric emission.—Electron emission from the cathode in a phototube, due to light or other radiant energy striking the cathode surface.

photoelectric material.—Metals such as barium, strontium, sodium, potassium, lithium, caesium or rubidium which emit electrons under the influence of light or other radiant energy.

photoelectric threshold.—The lowest frequency or longest wavelength at which it is possible for *photoelectric emission* to take place from the surface of a given cathode metal, this frequency or wavelength being a characteristic of the metal and varying with different metals. The frequency or wavelength at which the *quantum voltage* reaches the value of the *work function* voltage.

photo-electron.—An *electron* emitted from a cathode as a result of the action of *light*.

photoemissive.—Capable of emitting electrons under the influence of light or other radiant energy.

photoglow tube.—A gas filled phototube operated with voltage sufficiently great to cause ionization when illumination reaches the tube.

photolytic cell.—A photoelectric cell of the *photo-voltaic* type.

photomagnetism.—Effects resulting from the combined action of light and magnetism; or the science relating to such effects.

photometer.—A device for measuring intensity of *light* or for comparing the relative intensities of sources of light.

photometry.—Measurement of the visual effect of light.

photon.—1—A *positive ion*. 2—A *quantum* of energy.

phototransistor.—A transistor in which movement of electrons or holes is varied by changes of light on the junction of n-type and p-type crystal sections. Used similarly to a phototube.

phototube.—A vacuum tube or gaseous tube having for its cathode a substance which emits electrons when reached by light or other radiant energy. Emitted electrons are drawn to an anode which is maintained positive with respect to the cathode. These electrons flow from the anode through any external circuit and back to the cathode at constant or varying rate in accordance with constant or varying radiant energy on the cathode.

phototube sensitivity.—The ratio of changes of phototube current to changes of luminous or radiant flux which causes the current changes.

photo-voltaic.—Relating to the production of *electromotive force* by physical or chemical action in substances which are affected by *radiant energy*. A quality of certain electrolytic cells which may be used in a manner somewhat similar to photocells. A *cuprous oxide cell*. See illustration.

physical unit.—A unit based on natural or material values as opposed to values which are abstract or imaginary.

pi (π).—Greek letter symbol for 3.14159 . . ., a circle's circumference divided by its diameter.

pickup.—A *phonograph pickup*.

pickup circuit.—A *stand-by circuit*.

pie winding.—An inductance or choke coil formed by a number of sections one against the other, each of little width but of comparatively great outside diameter. *See illustration.*

Pierce oscillator.—A circuit commonly used in modified form for crystal controlled oscillators. It is somewhat similar to a Colpitts oscillator.

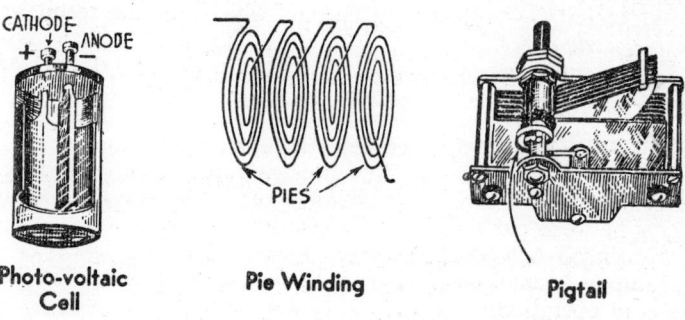

Photo-voltaic Cell   Pie Winding   Pigtail

piezo-electric crystal.—A mineral exhibiting the *piezo-electric effect*. The most common are quartz, rochelle salts and tourmalin.

piezo-electric effect.—The production of an *electrostatic charge* between opposite faces of certain crystalline substances (generally quartz) when they are compressed or twisted, and the opposite effect of a change in form of the same crystals when placed between electrostatic charges of opposite polarity. The mechanical extension and contraction, or vibration, of the crystal occurs most easily and energetically at certain frequencies determined by the crystal's physical dimensions, the action being analogous to electrical resonance in a tuned circuit.

piezo-electric loud speaker.—A loud speaker in which the mechanical movement for production of sound waves is obtained with a *piezo-electric crystal*.

piezo-electric oscillator.—A circuit containing a piezo-electric *crystal* maintained in a state of vibration by energy supplied from the circuit, the operating frequency being near one of the frequencies at which mechanical vibration of the crystal is naturally most energetic and easily maintained. The circuit in itself, without the crystal, is not capable of maintaining oscillations. The most common form places the crystal in the grid circuit of a vacuum tube.

piezo-electric pickup.—A phonograph pickup in which audio frequency currents result from varying mechanical pressure on a *piezo-electric crystal.*

piezo-electric resonator.—A body exhibiting the *piezo-electric effect;* having the ability to be set into resonant vibration by electrical means.

piezo-electric stabilizer.—A *piezo-electric crystal* or resonator used to fix the exact frequency at which oscillatory current is maintained in a circuit which is capable of maintaining oscillations within a narrow range of frequencies without the crystal. A *piezo-electric oscillator* circuit in which there is sufficient regeneration to maintain oscillation.

piezo-electricity.—Voltage and current which are the result of a *piezo-electric effect.*

piggy back antenna.—A television antenna with separate elements for high-band and for low-band reception mounted one above the other.

pigtail.—A length of wire permanently attached to and extending from a small fixed resistor, capacitor or other circuit component for convenience of connection and support. Also any flexible conductor between parts which have relative motion.

pile winding.—A *banked winding.*

pilot lamp.—A small lamp which is lighted while a piece of apparatus is in operation.

pincushion effect.—Inward bending of sides, top or bottom of television pictures, leaving corners extended outward.

pincushion magnets.—Small permanent magnets mounted near the flare of a television picture tube. Magnet bodies and poles may be varied in position to counteract pincushion effect.

pin jack.—A small receptacle into which is pushed a pin-like member to complete a circuit connected to the two parts.

pip marker.—A marker produced by voltage from a signal generator or oscillator and causing small extensions above and below some point on a frequency response trace.

pi-section.—A part of an electric circuit having two similar elements in parallel or across the circuit, and one element in series with one side of the circuit and connected between the two parallel elements. *See illustration.*

pitch.—The frequency of a *tone.*

pitch limit.—The sound frequency above which or below which the sense of hearing is no longer affected. See *auditory sensation area.*

plane-polarized light.—Light waves which vibrate (transversely to the line of propagation) in only one of the planes within which lies the line of propagation. *See illustration.*

plano-concave lens.—A lens having one plane or flat surface and another concave or depressed surface. *See illustration.*

plano-convex lens.—A lens having one flat or plane surface and another which is convex or outwardly bulging. *See illustration.*

plate.—In an electron tube of any kind, the element toward which electrons flow from the cathode while the plate is positive with respect to the cathode, and from which the electrons flow to and through external circuits on their way back to the cathode. An anode.

plate capacitance.—The sum of the separate *direct capacitances* between the plate of a tube and all the other elements in the tube. In a three-element tube, the sum of the *plate-filament capacitance* and the *grid-plate capacitance.* The symbol is C_p.

plate-cathode capacitance.—The *electrostatic capacity* between the plate and the cathode of a tube. The symbol is C_{pf}.

plate characteristic.—A graph showing for some certain type of tube the relations between plate current and plate voltage for any of a number of control grid voltages.

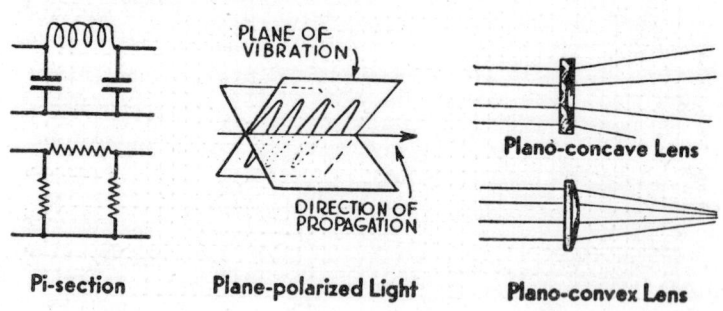

Pi-section   Plane-polarized Light   Plano-convex Lens

plate circuit.—The electrical paths followed by current flowing through the *plate* of a tube.

plate coil.—Any inductance coil forming part of the *plate circuit* of a tube.

plate condenser.—A condenser connected in the *plate circuit* of a tube.

plate conductance.—The ratio of a small change in *plate current* to the change in *plate voltage* which produces it, the *grid voltage* remaining constant.

plate current cutoff.—The condition with which control grid voltage becomes sufficiently negative to prevent flow of plate current in a tube.

plate current detection.—A method of *detection* in which the control grid of the detector tube has a *negative bias* sufficiently great that signal voltages are distorted. Positive swings of control grid voltage then result in changes of plate current that are greater than the changes resulting from equal negative grid voltages. These unequal plate current changes result in a rise and fall of average plate current corresponding to the signal *modulation frequency.*

plate current saturation.—The condition with which plate current in a tube does not increase to any great extent when the control grid is made less negative or positive. The result of small plate voltage.

plate detector.—A detector tube using the principle of *plate current detection*.

plate dissipation.—The power in *watts* which is used up in heating of the plates or anodes of tubes. Equal to the product of the *plate voltage* and *plate current*.

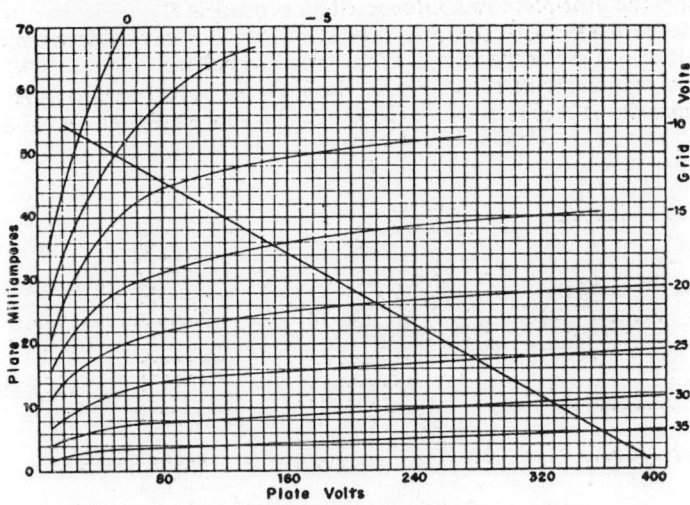

Plate Characteristics

plate-grid capacitance.—The *electrostatic capacity* existing between the plate and grid of a tube. The *grid-plate capacitance*.

plate impedance.—The impedance of the path between *plate* and *cathode* in a tube. Approximately the same as the *plate resistance*.

plate inductor.—A high impedance *choke coil* placed in a plate circuit to maintain a fairly constant value of direct plate current while the total plate current is changed by the varying grid voltages.

plate potential.—Plate voltage.

plate rectification detector.—A detector utilizing the principle of *plate current detection*.

plate resistance.—The resistance in ohms to flow of alternating currents between the *plate* and the *cathode* of a tube. The reciprocal of the *plate conductance*. The ratio of plate voltage changes to the resulting plate current changes in a vacuum tube, the grid voltage remaining constant. The symbol is R_p.

plate voltage.—The direct current potential difference between the *plate* and the *cathode* of a tube. In a D.C. filament tube the plate voltage is measured between the plate and the negative end of the filament. In an A. C. filament tube the plate voltage is the difference between the plate potential and the potential at a point midway between the ends of the filament. In an indirectly heated cathode tube the plate voltage is the potential difference between plate and cathode. The symbol for plate circuit voltage (at the source) is E_b, and for the true plate voltage as defined above the symbol is E_p or e_p.

plate winding.—A transformer winding connected in the *plate circuit* of a tube.

platinoid.—A resistance alloy composed of copper, nickel, zinc and a small amount of tungsten.

platinum.—A whitish, malleable metal little affected by high temperatures, by acids or by arcing on its surface.

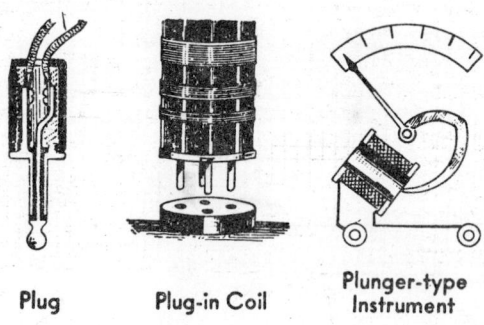

Plug   Plug-in Coil   Plunger-type Instrument

pliodynatron.—A four-element vacuum tube having the plate and one grid operating as in a *dynatron* and having the second grid operating as a *control grid* of the usual type, thus allowing operation as an oscillating detector with *negative resistance* characteristics to greatly strengthen the signal amplification.

pliotron.—A high-vacuum hot-cathode tube with one or more grids for controlling anode or plate current. A *triode, tetrode,* or *pentode.*

plug.—A metal sleeve and insulated tip inserted in a jack to complete a circuit. *See illustration.*

plug-in.—Descriptive of a part having its terminal connections made to metallic extensions or prongs which are slipped into corresponding openings or jacks to complete the circuit or circuits in which the device operates. *See illustration.*

plunger magnet.—A *solenoid* with a sliding core or plunger moved by magnetism in the solenoid.

plunger type instrument.—A *moving iron instrument* in which a long, slender piece of iron attached to the pointer is drawn into the core of a winding which carries the current to be measured. *See illustration.*

P$_m$.—Symbol for *maximum* power in watts.

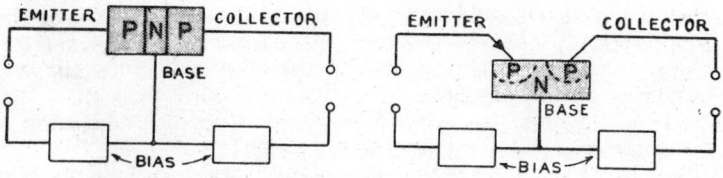

P-N-P Transistor

PM.—Abbreviation for permanent magnet.

p-n-p transistor.—A junction transistor in which emitter and collector are of p-type with between them a base of n-type.

point contact transistor.—A transistor in which the base is a crystal of either n-type or p-type on which rest emitter and collector consisting of finely pointed metallic conductors spaced a few thousandths of an inch apart. Emitter and collector are biased in opposite polarities, whether negative or positive depending on the type of crystal.

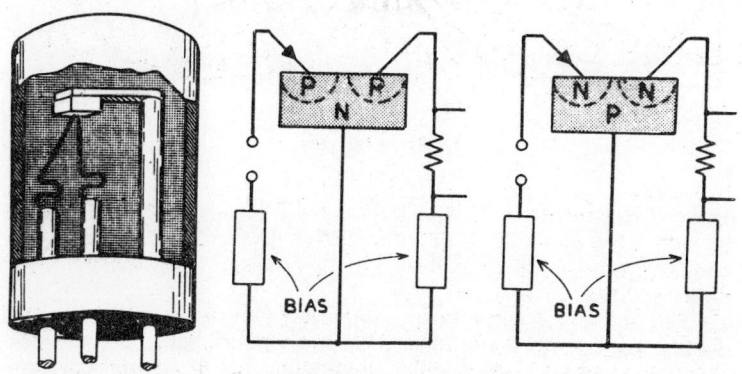

Point Contact Transistor

poisoning of cathode.—Presence of impurities or gases which reduce the electron emitting ability of a surface.

polar diagram.—A diagram showing pickup ability of an antenna for carrier signals approaching from various directions. On the

POLARITY

diagram are a number of lines radiating from a center at which is presumed to be the antenna, also concentric circles indicating percentages of maximum pickup. Curves or lobes show such percentages for all or selected directions of received signals.

polarity.—Descriptive of parts which are positive or negative in *potential* or in *electrification* when considered with reference to other similar parts at the opposite end of an electric or magnetic circuit. Having qualities at one end which are opposite or complementary to the similar properties at the other end.

polarization.—1—Formation of gas on the electrodes of a cell which is carrying an electric current, this gas producing a *counter- e.m.f.* which reduces the cell's effective voltage. Also, the difference between a cell's open circuit voltage and its working voltage as reduced by polarization. 2—A result of refraction, or reflection or of the effect of certain materials by which light waves are made to vibrate in definite directions or only in certain planes rather than vibrating in all directions which are transverse to the line of propagation. See *plane-polarized light*.

Polar Diagram

polarized capacitor.—A capacitor, usually electrolytic, which may be connected in a circuit carrying alternating or pulsating voltage and also direct voltage only with due respect to polarity of the direct voltage.

polarized light.—A light beam which has been subjected to *polarization*, so that all wave motion is in a single plane.

polarized magnet.—A magnet consisting of a current-carrying winding on a permanent magnet core, a small current then causing considerable change in the magnetic flux.

polarized relay.—A relay which operates upon flow of current in one direction but not with flow in the other direction.

polarized wave.—A carrier wave whose electrostatic lines are horizontal for horizontal polarization or vertical for vertical polarization. Television broadcast carriers are horizontally polarized.

polarizing unit.—Any electrical element, as a battery or a rectifier, which applies a steady direct potential to other parts, generally to parts such as microphones or loud speakers which are carrying alternating currents.

pole.—A terminal or a surface at which is exerted an *electromotive force* or at which an *electric current* or *magnetic lines of force* may enter or leave either a source or an energy consuming device. An *electrode*. A terminal of a source or a load. One end of a magnet.

pole changing switch.—A *double-pole, double-throw* switch arranged to connect both sides of one circuit to both sides of either one of two other circuits. *See illustration.*

polyphase.—More than one alternating electromotive force acting in a single circuit. *Two-phase* or *three-phase*.

porcelain.—A chinaware made from clay and feldspar; used for insulation. Dielectric constant 4.5.

port.—Any opening in a speaker baffle or cabinet other than that for the cone or radiator.

portable receiver.—A radio receiver with self-contained power supply, antenna system, amplifying stages and loud speaker, and sufficiently light in weight to allow carrying about.

portable transmitter.—A low power transmitter which is completely self-contained and designed to be moved from place to place.

positive.—Descriptive of any point in an electric circuit at which are fewer free electrons than at other (relatively negative) points in the same circuit, and toward which electrons tend to flow from such other points. Also a part positively charged due to having a deficiency of free electrons.

positive carrier.—A *positive ion*.

positive charge.—The quantity of electricity in a body which is at a positive potential with reference to another body carrying an equal negative charge.

positive column.—The luminous discharge existing near the anode in a gaseous tube in which *ionization* is taking place.

positive electrification.—The condition of a body on which there is a deficiency of negative electrons. See *electrification*.

positive image.—An image or picture in which lights and shades are the same in relative position as in the original.

positive ion.—An *atom* deficient in electrons, having a positive charge. An ion which moves toward a cathode. A *cation*.

positive modulation.—Television transmission in which increase of illumination of the picture image causes increase of radiation power.

positive nucleus.—That part of an *atom* which carries the positive charge, the negative charge being carried by the electrons which are associated with the nucleus in the atom.

positive reactance.—*Inductive reactance*.

positive temperature coefficient.—A characteristic of a substance in which the *resistance* increases with rise of temperature. True of all metals and most metallic alloys.

positive transmission.—Television carrier modulation with brightest areas of pictures represented by maximum amplitudes while tips of sync pulses are at minimum amplitudes. Not used in the United States.

post deflection focusing.—In a single-gun color television picture tube, focusing of the electron beam, after vertical and horizontal deflection, onto either blue, green or red phosphor strips by means of charged conductors in a color grid immediately back of the phosphor plate.

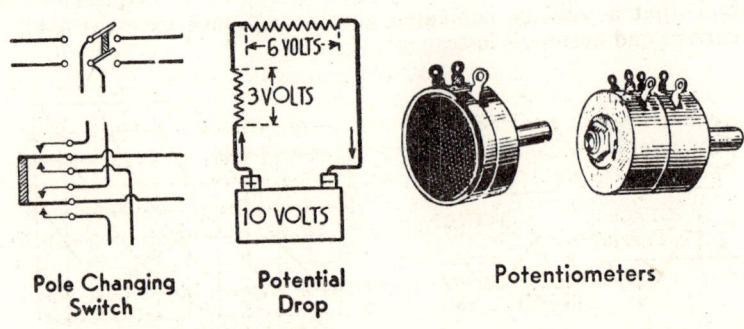

Pole Changing Switch

Potential Drop

Potentiometers

potential.—A measure of the relative voltage, the amount of electric charge or the degree of electrification at a point in either an electric circuit or an electric field when that point is considered with reference to some other point in the same circuit or field. Measured in *volts*.

potential difference or drop.—The difference in electric pressure which causes movement of electricity from a point of higher pressure to one of lower pressure or potential. Measured in *volts*. The symbol is V. Abbreviated *p.d.* *See illustration*.

potential divider.—A *voltage divider*.

potential energy.—The energy or working ability contained in anything because of its position or shape. An electric charge on the plates of a condenser represents potential energy. Compare *kinetic energy*.

potential gradient.—1—The rate at which the *potential* decreases within the space between two points. The number of volts change in a given distance, divided by the distance. The maximum possible potential gradient is that of the material's *dielectric strength*.
2—The *electrostatic field intensity*.

potential winding.—A winding in parallel with a circuit, connected across the two sides of the circuit.

POTENTIOMETER

potentiometer.—A device providing adjustable resistance or else voltage divider action. The most common form consists of a circular or cylindrical resistance element of either composition or wire-wound type on which may be rotated a contacting slider. There are terminals for the slider, for opposite ends of the element and sometimes for intermediate points. Also an instrument for measuring an e.m.f. by opposing with another known e.m.f., usually with means for determining the ratio of known to unknown values.

Potter oscillator.—A cathode coupled multivibrator of the type often used in television sweep oscillators.

powdered iron core.—A magnetic core consisting of finely divided iron whose particles are held together but insulated from one another by cement. Used in inductors operating at frequencies so high that a solid or laminated core would have excessive eddy current and hysteresis losses.

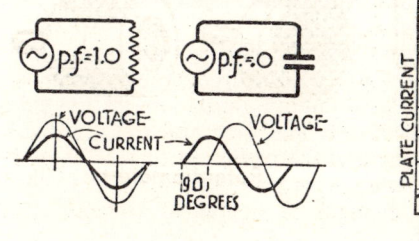

Power Factor

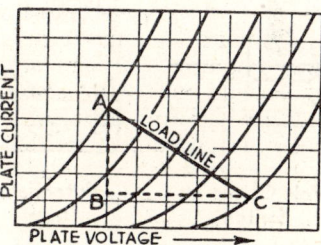

Power Triangle

power.—The rate at which work is done or at which energy is used: as by movement of a given weight through a given distance within a given time. Electric power is measured in *watts*. The symbol for average power in watts is P, for instantaneous power it is p., and for maximum or peak power it is P_m.

power amplification.—The ratio of the alternating output power to the alternating input power.

power amplifier.—1—An audio frequency amplifier with one or more power tubes in its output stage. 2—A radio frequency amplifier between oscillator and antenna in a transmitter.

power factor.—1—A measure of the ability of a circuit or device to turn alternating voltage and current into useful power or to dissipate power. The greatest possible ability corresponds to a power factor of 1.0 (unity) and inability to produce or dissipate any real power corresponds to a power factor of 0.0 (zero). The power factor is equal to the ratio of the actual power in *watts* to the apparent power in *volt-amperes*, or is equal to the cosine of the angle by which the current either lags or leads the applied voltage. 2—Power factor of a coil is equal to the coil's *resistance* divided by its *impedance*, or to the cosine of the angle of lag. 3—Power factor of a condenser is equal to the condenser's *resistance* divided by its *capacitive reactance*, or to the cosine of the angle of lead. Abbreviated *p. f. See illustration.*

power factor correction.—The addition of condensers to an *inductive circuit* to bring the current and voltage more nearly in phase and thus to increase the proportion of volt-amperes turned into real power or to increase the *power factor.*

power level indicator.—A volume indicator calibrated to read in units of power, or in power ratios expressed in *transmission units, decibels,* etc.

power stage.—A stage of *audio frequency* amplification in which a power tube is employed. Usually the final stage in an amplifier, from which power goes to the loud speaker.

power transformer.—A transformer which produces from line power the alternating currents for filaments and heaters, also the currents which are rectified and filtered for the plate, screen and grid circuits in radio apparatus.

power triangle.—A right angled triangle constructed on a series of vacuum tube *plate characteristics;* the hypotenuse being the *load line* between permissible grid voltage limits, one of the sides being a horizontal projection of this load line and the other side joining the ends of the load line and the projected line. The product of the two sides of this triangle is proportional to the tube's relative power output with changes of operating voltages. *See illustration.*

power tube.—A receiver amplifying tube having a *maximum undistorted power output* in excess of one-half watt and generally having a low amplification coefficient. A tube capable of amplifying a large signal voltage without distortion.

power unit.—All of the apparatus which furnishes filament and cathode heater current, plate current, screen current and bias voltages for radio receivers and other electrical devices using commercial line power, either alternating or direct current, for the primary source of energy. The power unit includes a *filter* and a *voltage divider,* and if used with A.C. line supply, it also includes a *power transformer* and a *rectifier. See illustration.*

practical unit.—A unit of convenient value, a unit equal to certain multiples or certain portions of the *absolute units.* Practical units include the *ohm, volt, ampere, farad, henry, watt,* etc.

preamplifier.—1—A voltage amplifier used between the input of an audio amplifier and any source of weak signal voltages. 2—A booster for television or radio receivers.

pre-emphasis.—In frequency-modulation sound transmission, increase of amplitude at the higher modulating frequencies for the purpose of overcoming noise voltages which otherwise might cause disagreeable hiss. A similar method employed in sound recording.

PREFERRED VALUES

preferred values.—A series of values for resistors and capacitors such that the increase between any two steps is of the same percentage as between all other steps. Increases may be in steps of 20, 10 or 5 per cent each.

press.—The glass support which carries supporting wires and connections for the elements of a tube. *See illustration.*

pressure.—The amount of force acting on a given area or space. Electric pressure is called *electromotive force, voltage* or *potential*.

primary cell.—A combination of electrodes and chemicals which produce an *electromotive force* at the expense of changes in the chemical composition of the parts. The changes are such as cannot be reversed for restoration of the cell upon exhaustion. A cell which must be replaced or in which the elements must be replaced after the production of a certain amount of electrical energy, as

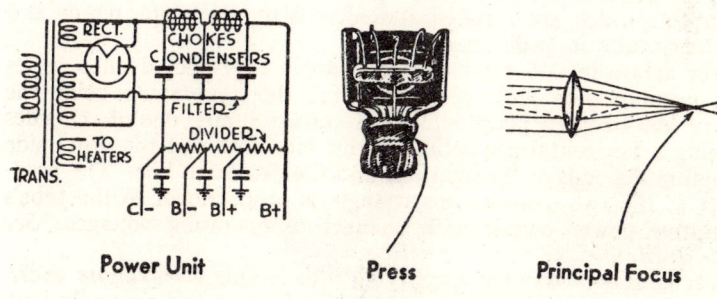

Power Unit   **Press**   **Principal Focus**

distinguished from a *storage cell* or secondary cell in which a flow of current from an external source restores the elements to their original forms.

primary circuit.—The circuit in which appears the applied electromotive force; as the circuit through which power enters a transformer, or the circuit taking power from a source and delivering it to the load or to a secondary circuit.

primary colors.—In color television, three bands of visible wavelengths centering at blue, green and red of the color spectrum. They may be combined in various proportions to give the visual impressions of all other colors as well as of white.

primary electrons.—Electrons resulting from *primary emission*.

primary emission.—The *electron emission* due to primary causes such as heating of a cathode, and not to secondary effects such as *ionization* or *electron bombardment*.

primary winding.—A winding connected in a *primary circuit*.

principal focus.—The focus for rays which are parallel before being acted upon by a lens. Also the *virtual focus* for parallel rays acted upon by a diverging lens. *See illustration.*

PRINTED WIRING

printed wiring.—Circuit conductors consisting of thin copper or other conductor of various lengths and shapes as may be required, securely bonded to a base of insulation. The conductors are formed to shape by chemically etching away the unneeded conductor from an original continuous coating. Circuit components usually are mounted on the base material and connected to terminal posts or eyelets on the printed conductors.

probe.—A device mounted on the free end of an instrument cable for making suitable connections to circuits being measured or tested. A probe may contain resistors for isolating the cable capacitance from tested circuits, or it may contain a matching pad, a detector for demodulating high-frequency carrier voltages, or other elements required for any particular function or application.

projection television.—A method of television picture reproduction with which strong illumination from a picture tube operated with high anode voltage passes through a system of mirrors and lenses and is thrown on a large viewing screen which is part of the receiver or is externally mounted.

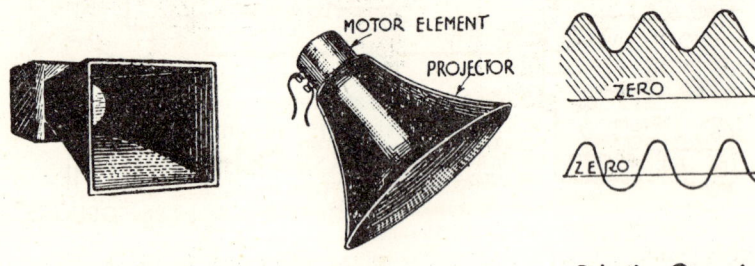

Projectors   Pulsating Current

projector.—The sound radiating portion of a loud speaker. A horn, baffle or the like. *See illustration.*

propagation.—Passage of a disturbance through a medium; such as the passage of radio waves through the ether, passage of sound waves through air, etc.

propagation constant.—In a transmission network or line consisting of an infinite number of similar sections, the natural logarithm of the vector ratio of the current entering one such section to the current leaving it.

protective device.—Any device which keeps dangerously large currents out of a circuit.

psi (ψ).—Greek letter symbol for *dielectric flux*.

p-type crystal.—A transistor crystal in which are impurities whose atoms have fewer valence electrons than atoms of the pure crystal. Resulting points in the crystal which lack a normal electron form so-called holes.

PUBLIC ADDRESS SYSTEM

public address system.—Electrical and acoustic apparatus used for reproduction of speech and music with volume and tone quality suitable for large audiences or large spaces.

pulling.—Bending in television pictures.

pulsating current.—A direct current which rises and falls regularly. A current equivalent to a *direct current* and an *alternating current* in one circuit. The *amplitudes* in one polarity are greater than those in the opposite polarity so that the average value is not zero.

pulse.—A momentary flow of electricity.

puncture voltage.—*Dielectric strength.*

pup jack.—A small receptacle into which fits a single contact plug to join two conductors. A *tip jack*.

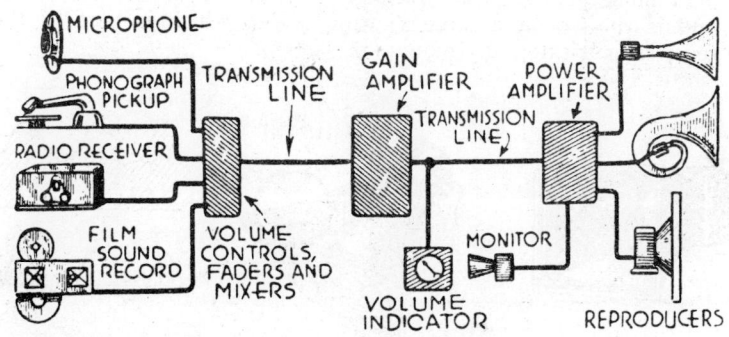

pure inductance.—An inductance assumed as existing without capacity or resistance in the same element. See *apparent inductance*.

pure resistance.—A resistance with which is associated neither capacity nor inductance in the same unit.

pure tone or note.—A sound produced by waves of a single frequency, without *harmonic frequencies*.

pure wave.—A wave in which nearly all the energy is concentrated at one sharply defined frequency.

purity.—A characteristic of color television pictures which exists when the electron beam or beams in the picture tube strike phosphor dots or strips of only one primary color at a time, thus allowing correct degrees of saturation and no contamination of one primary color with others.

purity coil.—An electromagnet outside the neck of a color television picture tube to assist in directing electron beams to the proper phosphor dots by adjustment of coil position and current.

PURITY MAGNET

purity magnet.—One or a group of permanent magnets adjustably mounted on the neck of a color television picture tube so that the combined magnetic fields allow obtaining purity in pictures.

push-pull amplification.—An amplification system employing two tubes or two sets of tube elements whose control grids are fed simultaneously with signal voltages of opposite phase obtained from opposite ends of an input transformer secondary or else from the input signal and the output of an inverter. Resulting plate signal currents, also in opposite phase, are fed to the outer ends of the primary in an output transformer or to an equivalent arrangement of chokes. Plate signal components add in the output while variations of direct plate current cancel.

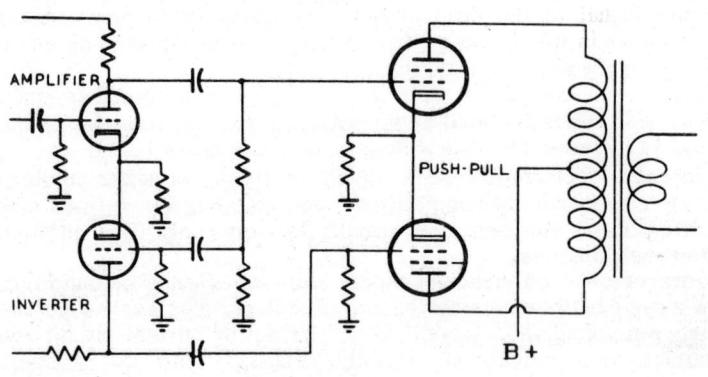

Push-pull Amplifier

push-pull oscillator.—A vacuum tube oscillator using two tubes with their control grids and plates connected to opposite ends of tuned circuits, the operation being similar to that in *push-pull amplification*. Each tube works on alternate half cycles of the high frequency currents.

push-pull transformer.—A transformer having a center tapped secondary, a center tapped primary, or both and designed for use in a *push-pull amplification*.

Q

Q.—Symbol for quantity of electricity in *coulombs* or in *ampere-hours*.

Q-demodulator.—A color television demodulator in which combine the chrominance signal and a color oscillator output voltage to produce a Q-signal.

Q-factor.—A figure of merit for a capacitor, inductor, or entire circuit. Equal to the ratio of ohms of reactance to ohms of that resistance in which energy dissipation would be equal to all energy losses in the element or circuit considered.

Q-signal.—A component of a color television chrominance signal, with side-bands centered at 90 electrical degrees from the I-signal and 147 degrees from the reference or burst phase.

Q-signals.—Telegraphic *code* signals consisting of letter combinations always commencing with Q, and standing for various complete phrases and sentences, usually for commonly used questions and their answers.

quadrature.—A difference of ninety *electrical degrees* or one-fourth of a cycle between alternating current values. The *phase difference* between voltage and current in an alternating current circuit containing only inductance or only capacity, with no resistance present.

quadrature component.—The *reactive component*.

quality.—*Fidelity*, especially with reference to *tone*.

quantity.—Quantity of electricity, measured in *coulombs* or *ampere-hours*.

quantum.—A unit of *energy*.

quantum theory.—A theory of radiation or emission stating that the radiating or emitting body does not emit energy uninterruptedly but rather intermittently in units called quanta (quantums), the value of which is dependent on the operating frequency of the radiating body and on a constant.

quantum voltage.—The *energy*, expressed in volts, which is acquired by an electron from light of a given *wavelength;* the voltage varying inversely with the wavelength.

quarter phase.—*Two-phase*.

quarter-wave line.—A resonant line whose electrical length is equal to one-fourth the wavelength corresponding to frequency of resonance in the line.

quartz.—A natural mineral forming an excellent insulator with low radio frequency losses. Dielectric constant 4.5. The most

QUARTZ CRYSTAL OSCILLATOR

important piezo-electric crystal. Quartz is transparent to ultraviolet light, to which ordinary glass is opaque.

quartz crystal oscillator.—A *piezo-electric oscillator.*
quartz plate.—A *piezo-electric crystal* made from quartz.
quartz resonator.—A *piezo-electric resonator.*
quiescent aerial radiation.—Radio telephony in which the *carrier wave* is radiated only while *modulation* takes place or while signals are being sent.

R

R. or r.—Symbol for *resistance* in ohms.

radian.—An angle formed at the center of a circle by lines drawn to the ends of an arc which has a length equal to the circle's radius. This angle is equal to 360 degrees divided by 2π, which is 57° 17′ 44.8+″, or about 57.2958 degrees. The word sometimes is used as meaning the arc included by this angle. *See illustration, page following.*

radiant energy.—Any form of energy which may be sent through space by means of *electromagnetic waves*. Light, radiant heat and radio waves are forms of radiant energy.

radiant flux.—The rate of flow of *radiation* with respect to energy. Measured in watts or in ergs per second.

radiant heat.—Heat energy carried through space by wave motion at frequencies lower than those of visible light, the waves manifesting themselves by raising the temperature of matter in their path.

radiate.—To send *electromagnetic waves* into space.

radiating system.—An *antenna* circuit for *radiation* of signals.

radiation.—Transfer of *radiant energy* through space by wave motion. Specifically, the action by which *radio waves* are produced and sent into space by the antenna system of a transmitter.

radiation field.—The force which detaches itself from an antenna or other circuit carrying *oscillatory current* and travels through space, this force resulting from combinations of the *magnetic* and *electrostatic fields* set up by the transmitting antenna system or other circuit.

radiation power.—The power in *watts* radiated from an antenna; equal to the square of the maximum effective *antenna current* multiplied by the *radiation resistance*.

radiation resistance.—The value of *resistance* which, inserted at a point of maximum current in an antenna system, would consume the same amount of power that leaves the antenna through *radiation*. The ratio of the total *radiation power* to the square of the maximum effective current in the antenna.

radiator.—1—An *acoustic radiator*. 2—An antenna used for *radiation* of signals.

radio.—1—The art of communicating and reproducing intelligence of any form by means of *wave motion* radiated through space or by means of carrier currents in conductors. Also, that part of electrical science relating to this art. 2—A *radio receiver*. 3—A message sent by radio, or the act of sending such a message.

radioactive rays.—Rays capable of penetrating materials which are opaque to ordinary light. *X-rays* and other rays of short wavelength.

radioactive substance.—A substance which emits waves capable of penetrating materials through which *visible light* does not pass. A substance from which electrically charged particles are emitted upon breaking down of its atoms. Among such substances are the elements *radium, thorium* and *uranium*.

radio beacon.—A stationary radio transmitter sending special signals which allow mobile receivers in a ship, an airplane, etc., to determine their direction and course with reference to the transmitter.

radio channel.—The band of frequencies within which the *modulated wave* of a transmitter is allowed to operate. A band of frequencies sufficient in extent to allow transmission of some form of signal without causing interference above a certain intensity in frequencies outside the band.

radio compass.—A *mobile receiver* allowing determination of the direction from which a radio signal is coming to the receiver. A *direction finder* used in navigation.

radio field.—A *radiation field*.

radio field intensity.—The effective value of the electric or magnetic field of a signal at a given point, as expressed in number of *microvolts or millivolts per meter* of height of a receiving antenna.

radio frequency.—A frequency sufficiently high to allow effective *radiation*. By definition, a frequency of thirty kilocycles or more. A term sometimes used to mean any frequency above the limits of normal audibility, although some frequencies which are audible also can cause radiation. Abbreviated *R.F., r.f.* or *r-f*.

radio frequency amplification.—An increase in the voltage or power of a signal at *radio frequencies*. The amplification between the antenna and first detector of a receiver.

radio frequency amplifier.—Vacuum tubes and coupling circuits which increase signal voltage or power at *radio frequencies*.

radio frequency choke.—An air-core *inductance coil* providing high impedance at radio frequencies. A *high frequency choke*.

radio frequency coil.—An air-core *inductance coil* designed to operate at high frequencies without excessive loss.

radio frequency oscillator.—A device including one or more *tuned circuits* in which are produced radio frequency currents and from which may be secured current at radio frequency or *radiation fields*. The circuits may be excited by means of vacuum tubes, a buzzer or other suitable devices. See under names of oscillators: *Hartley, Colpitts, Meissner, tuned-grid tuned-plate*, etc.

radio frequency resistance.—*High frequency resistance*.

radio frequency selectivity.—The ratio of a circuit's effective voltage or current at *resonance* to the effective voltage or current when the circuit is detuned by one per cent of the *resonance frequency*.

radio frequency transformer.—A transformer designed for operation at *radio frequencies*, generally having an air core but sometimes having a small amount of iron in its core. *See illustration, page following.*

radio frequency tube.—A vacuum tube especially designed for use in radio frequency circuits, having low inter-electrode capacity. Any tube used in apparatus working at *radio frequency*.

radiogram.—Message transmitted by *radio telegraphy* or *telephony*.

radiometer.—A device consisting of rotating vanes in a gaseous or evacuated chamber, used to indicate the intensity of *radiant energy*. *See illustration*

radio noise field intensity.—The field intensity of the electromagnetic waves which produce *interference* at a given point.

radiophone.—1—Pertaining to *radio telephony*. 2—Apparatus for production of sound from radio signals.

radio range.—A *radio beacon* sending out directed waves from which a receiver in motion may note any deviation from a given course of travel.

radio receiver.—Apparatus including *tuned circuits* and a *detector*, with or without *radio frequency amplification* and *audio frequency amplification*, used to produce audible sounds from *modulated wave* signals.

radio relay station.—A radio station which receives messages from

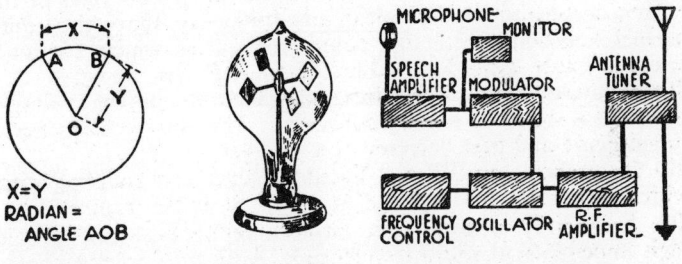

Radian   Radiometer   Radio Transmitter

another and transmits them to a third station which is their final destination or which is nearer to the destination.

radioscope.—A device showing the effect of *radioactive rays*. A *fluoroscope*.

radio spectrum.—1—All of the frequencies included within the limits of *radiant energy*. 2—The frequencies employed in one certain class of radio communication.

radio telegraphy.—Radio communication by means of *code* signals.

radio telephony.—Radio transmission and reception by means of *carrier waves* modulated by speech, music and other sounds.

radiotherapy.—Treatment of physical disorders by rays from *radioactive substances*, by *ultra-violet rays*, etc.

radio transmission.—Transmission of signals by *radio waves* originating at a circuit especially arranged for such work.

radio transmitter.—Apparatus for production and signal *modulation* of radio frequency power. Oscillating, amplifiying and modulating circuits with all their associated apparatus, operating to energize an antenna system for *radiation* of radio waves into space. *See illustration.*

radiotron.—A trade name for *vacuum tubes.*

radio wave.—An *electromagnetic wave* capable of carrying signal modulation.

random winding.—A coil winding in which the turns and layers are not regularly positioned or spaced.

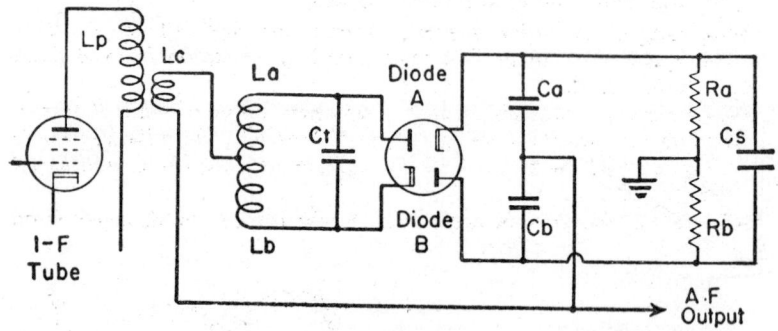

Ratio Detector

raster.—Illumination of a television picture tube viewing screen due to horizontal and vertical deflection of the electron beam without variations of beam intensity by picture signals.

ratio detector.—A demodulator for frequency-modulated signals. Primary and secondary of a transformer are made resonant at the center frequency. With no deviation, voltage at one end of the secondary leads primary voltage by 90 degrees and at the other end lags by 90 degrees. These voltages go to the plate of one diode and cathode of another. Primary voltage phase is taken through a third winding to a secondary center tap and a center tap on the other side of the diode circuit, thus being applied equally to both diodes. Frequency deviation brings one secondary voltage more nearly in phase with the primary while shifting the other secondary voltage further out of phase. The combined voltages increase conduction in one diode while decreasing it in the other, causing proportional changes of rectified voltage across diode load resistors. The ratio of resistor voltages varies with deviation and becomes audio output. The sum of these voltages is held nearly constant during audio variations by a large capacitance which absorbs or limits any amplitude modulation in the f-m signals.

Rayleigh disc.—A means for measuring the air pressure due to a *sound wave*, a light metal disc being suspended in the path of the wave so that the disc is deflected proportionately to the intensity of the sound.

reactance.—In an alternating current circuit, that part of the opposition to current flow which is due to inductance, to capacity, or to both inductance and capacity in the circuit. That part of the impedance which results from self-inductance and capacity. *Inductive reactance, capacitive reactance*, or the net effect of both. Reactance is measured in *ohms*. The symbol is X or x. See *illustration*,

reactance coil.—A *choke coil*.

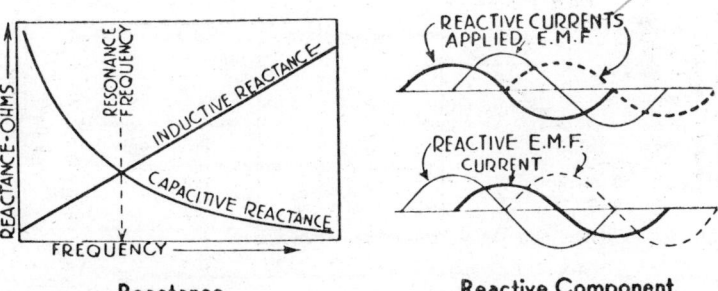

Reactance   Reactive Component

reactance tube. A tube connected across the tuned circuit of an oscillator to vary effective inductive or capacitive reactance and thereby return the oscillator circuit to resonance when its frequency tends to vary with respect to a synchronizing frequency. Change of frequency in the reactance tube cathode circuit makes its plate current either lag or lead plate voltage and oscillator voltage, depending on whether frequency increases or decreases. The value of lagging or leading plate current and its reactive effect in the oscillator circuit may be made greater or less by voltage applied to the reactance tube grid from any automatic frequency control system.

reactance variation method.—A method of measuring *high frequency resistance* according to the change in current caused by a known alteration of the *reactance* in a resonant circuit.

reaction.—1—The mutual effects of two quantities upon each other. With one body being affected by a force from a second, reaction is the opposite force exerted by the second body upon the first one. 2—*Regeneration*.

REACTIVE CIRCUIT

reactive circuit.—A circuit in which the *inductive reactance, the capacitive reactance* or both are of comparatively high value with respect to the circuit's *resistance.*

reactive current or component.—In an alternating current circuit, the portion of the current which is ninety degrees out of phase with the applied voltage, or the voltage which is ninety degrees out of phase with the current it produces. The current or voltage which charges the *capacity* or produces a field around the *inductance* in a circuit, from both of which energy is returned to the circuit so that this reactive component does no useful work and dissipates no *power.* The wattless current or component. Compare *active current or component. See illustration.*

reactive drop.—The drop in voltage due to *reactance* in a circuit.

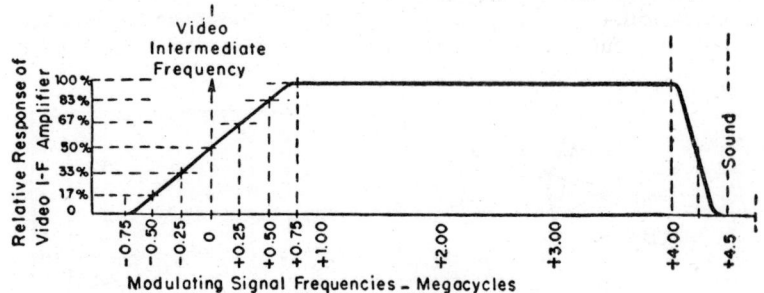

Receiver Attenuation

reactive load.—A load in which the current is not in phase with the applied voltage, either leading or lagging the voltage. A *capacitive load* or an *inductive load.*

reactor.—A part providing *inductive reactance* for operation, control or protection of electrical circuits and apparatus.

real component.—The *active component.*

real focus.—The position at which an image is formed by a lens.

real image.—An image which may be focused upon a screen, an image which is formed by the light rays themselves after *refraction* or *reflection,* and which always is inverted. *See illustration.*

receiver.—1—A *radio receiver.* 2—A *telephone receiver.*

receiver attenuation.—In television receivers, the form of i-f amplifier frequency response that compensates for doubly transmitted and received frequencies in vestigial sideband transmission.

receiver response.—1—The ratio of sound intensity or pressure in bars on a conveying medium (such as air) at a specified distance from the source to the applied signal voltage acting across a suitable resistance connected to the receiver terminals. The "receiver" is a loud speaker. 2—The receiver output in *decibels* when using a reference level resulting from the condition of 1-volt signal across 1-ohm resistance with a 1-bar sound pressure.

reciprocal.—In the most common use of the word, the reciprocal of a quantity is equal to 1 divided by the quantity.

reciprocal ohm.—A unit of *conductivity*, the reciprocal of one ohm. One *mho*.

recording head.—The portion of a tape or wire recorder containing an electro-magnet that magnetizes the tape or wire as it passes the head. The head may be designed to allow recording, reproduction, or erasing.

recording tape.—Tape made with plastic or paper base on which is a layer of some form of iron which may be magnetized in accordance with variations of recorded sound. The tape may be demagnetized, then used over again.

rectification.—1—The process of changing *alternating current* into pulsating current or into a current having a direct component.

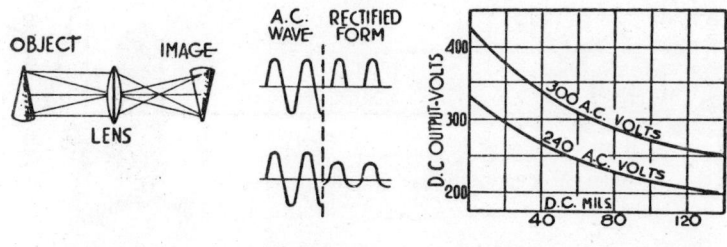

Real Image   Rectification   Rectification Characteristic

See illustration. 2—A name applied to the process of *detection*.

rectification characteristic.—A series of curves showing the relations between direct current, direct current voltage and applied alternating voltage in a *rectifier*. *See illustration.*

rectification factor.—The ratio of change of *direct current* in a rectifier circuit to the change of applied *alternating* voltage, other voltages remaining constant.

rectified current.—A current having a *direct component*, as delivered from a *rectifier*.

rectified signals.—Signals at audio frequency which result from passing a modulated carrier through a *detector* system.

rectifier.—Any device in whose output is direct or one-way current when alternating voltage is applied to the input.

rectifier instrument.—An instrument for the measurement of *alternating* voltages or currents, the alternating current being rectified in a *bridge rectifier* consisting of contact rectifier elements, and the resulting direct current operating a *moving coil instrument*. *See illustration.*

rectifier tube.—A *thermionic rectifier* or a *gaseous conduction rectifier*.

rectigon rectifier.—A trade name for an *argon rectifier*.

rectilinear lens.—A lens in which the effects of *aberration* are prevented, all lines being correctly focused.

red gun.—In a three-gun color television picture tube, the electron gun whose beam is intended to excite only the phosphor dots that emit red light.

reference electrode.—Any electrode or element in a tube or transistor from which voltages are measured to other electrodes.

reference level.—For measurements in decibels, the value of load and of power into that load assumed to represent zero decibels. It is the zero level or starting point for calculations of gain or loss in power, voltage, current, or other quantity.

reference phase.—Phase of the burst transmitted with color television carriers for use in synchronizing the receiver color oscillator with transmitted color signals.

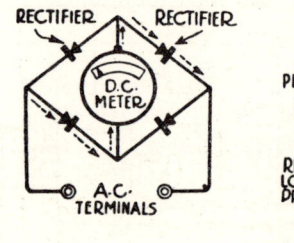
Rectifier Instrument

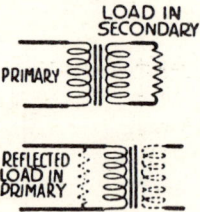
Reflected Value

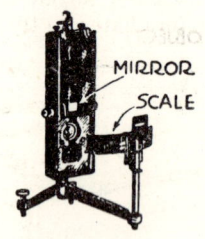
Reflecting Galvanometer

referred value.—A *reflected value*.

reflected value.—The effective or apparent value of a quantity or electrical effect as it appears to exist in one circuit, this value being due to real quantities or effects in a coupled or connected circuit. *See illustration*

reflected wave.—1—A carrier wave that reaches a receiving antenna only after being reflected from the surface of some good or fairly good electrical conductor. 2—That portion of any wave which is turned back in a circuit, as distinguished from the portion absorbed or passing onward. 3—The sky wave in space radio.

reflecting galvanometer.—A galvanometer to the moving part of which is attached a mirror for reflection of a beam of light onto a scale or for reflection of the image of a scale into a telescope. *See illustration.*

reflection.—1—The turning back from a surface of light, heat, sound, radio or other wave motions reaching the surface. Radio waves are reflected from a conductive medium. 2—In transmission lines, a result of incorrect terminal impedance relations because of which certain frequencies cause *nodes* and *antinodes* of voltage and current, or *standing waves*, at various points along the line.

reflection altimeter.—A device for determining an aircraft's height above the ground by the effect on an oscillator's frequency of *phase difference* between a radiated wave and the wave which is received after being reflected back from the ground below. Determination also may be made by measurement of *beat frequencies* resulting from reflected waves produced at two different radiated frequencies.

reflection coefficient.—The percentage of light or other energy of wave motion which is reflected back from a surface reached by the energy.

reflection loss.—Energy loss due to an incorrect *impedance match*, the loss caused by *reflection* effects in transmission lines.

reflector.—One or more conductors mounted parallel to a diode antenna on the side opposite to that approached by desired signals. Carrier signal energy is picked up by the reflector and part is reradiated to the dipole.

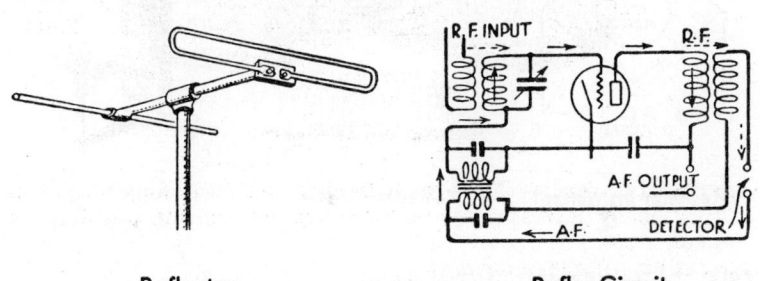

Reflector   Reflex Circuit

reflex circuit.—A vacuum tube receiving system in which one tube is made to amplify at both *radio frequency* and *audio frequency*, or is made to operate both as an amplifier and a detector, or is made to perform all three functions at the same time. *See illustration.*

refraction.—A bending or change in the direction of travel of light, heat, sound or radio waves which takes place as they pass obliquely from one medium into another or pass between portions having different characteristics in a medium. The refraction takes place at the boundary where there is a change in rate of penetration or in speed of the waves. *See illustration, page following.*

refraction index.—The *index of refraction*.

regeneration.—A method of increasing the total energy input to any amplifying system by returning to the input a part of the

REGENERATIVE CIRCUIT

energy appearing in the output. In a radio frequency amplifier, a vacuum tube detector or an audio frequency amplifier, a *feedback* of plate circuit voltage to a preceding grid circuit in such phase relation that the grid voltage changes are increased in amplitude.

regenerative circuit.—A circuit in which *regeneration* occurs.

regenerative coupling.—A coupling through which there is an energy feedback to cause *regeneration*. In amplifying tube circuits this coupling may be of any type and external to the tube, or it may be through the plate-grid capacitance inside the tube.

regenerative detector.—A vacuum tube detector, usually of the grid current detection type, with which a *feedback* of radio frequency energy from plate circuit to control grid circuit produces *regeneration*.

regenerative loop.—A receiving *loop antenna* in which part of the turns carry output current or plate current and provide a *feedback* of energy to the turns which are connected across the input or grid circuit of an amplifying tube thus allowing *regeneration*.

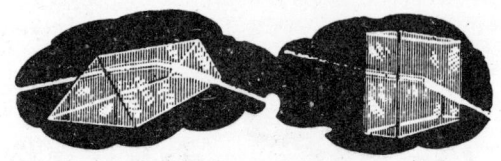

Refraction of Light

regenerative receiver.—A radio receiver in which amplification is intentionally increased by *regeneration*, generally in the detector circuit.

regulation.—*Voltage regulation*.

regulation curve.—A graph showing the relation between voltage and current in an electrical system, a graph showing *voltage regulation*.

regulator tube.—A *voltage regulator tube*.

rejector circuit or rejector resonance.—A *parallel resonance* circuit tuned to a frequency to be kept out of a second circuit with which the first one is in series, the parallel resonance circuit offering maximum impedance to series currents of the tuned frequency.

relaxation oscillator.—Any oscillator whose operating frequency is determined by time required for a capacitance to charge and discharge through a resistance.

relay.—A device which employs electromagnets, vacuum tubes, gridglow tubes or other means to control the current or alter the connections in one circuit when there are changes in another controlling circuit or instrument, the relay being connected to both parts. Compare *vacuum tube relay*. *See illustration*.

relay station.—1—A point in a transmission line at which are installed *telephone repeaters, attenuation equalizers* and other control equipment. 2—A *radio relay station*.

reluctance.—Opposition to the passage of *magnetic lines of force* in a magnetic circuit. Similar to resistance in electric circuits. Measured in *oersteds*. Reluctance in oersteds is equal to *magnetomotive force* in gilberts divided by the *flux* in lines. The symbol is the script letter "*R*".

reluctivity.—Specific reluctance; the *reluctance* between opposite faces of a centimeter cube of a substance. The reciprocal of *permeability*.

remanence.—*Residual magnetism*.

remote control.—Electrical or mechanical control of tuning, volume, switching and other functions of a radio receiver or other device from a point distant from the device.

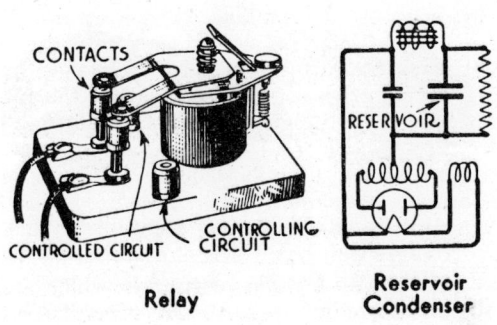

Relay

Reservoir Condenser

remote cutoff.—Descriptive of a tube in which plate current is reduced to zero only by making the control grid strongly negative, often five or more times as negative to the cathode as with an otherwise similar sharp cutoff tube.

repulsion instrument.—A current measuring instrument employing the repulsion effect between two similarly magnetized iron poles or between the fields of two windings to move the indicating pointer. A *vane instrument*.

re-radiation.—1—Radiation from a receiver's antenna of a signal which has been received and so greatly amplified by *regeneration* as to make such radiation possible. 2—Radiation from a receiver's antenna of unmodulated waves generated by oscillating receiver circuits which are coupled to the antenna.

reservoir condenser.—In a *power unit* filter, the condenser connected across the filter output and farthest from the rectifier. *See illustration*.

residual capacity.—A name sometimes applied to the small capacity between plate and control grid of a *screen grid tube*.

residual charge.—The electric charge which remains in a condenser after the initial discharge, and which will cause a second smaller discharge. The charge represented by flow of *absorption current*.

residual magnetism.—The magnetism which remains in a piece of iron or steel after the *magnetizing force* has dropped to zero.

resistance.—1—Opposition to flow of either direct or alternating electric current. The opposition which results in production of heat in the material carrying the current. Resistance increases directly with the length and inversely with the cross section of a conductor; it increases with rise of temperature in all metals and most metallic alloys and decreases with rise of temperature in carbon and most liquids. *Ohmic resistance*. Measured in *ohms*. The symbols are R, r, or the Greek letter omega (ω or Ω). 2—A *resistor*.

resistance box.—A number of resistance elements so arranged that various combinations may be used in a circuit.

resistance bridge.—A *Wheatstone bridge* arranged to measure resistances.

resistance-capacity coupling.—A coupling method by which voltage changes developed across a *resistance* in the plate circuit of one tube are applied to the grid circuit of a following tube through the *electrostatic capacity* of a condenser. *See illustration*.

resistance coupled amplifier.—Generally an amplifier utilizing *resistance-capacity coupling*.

resistance coupling.—A transfer of energy from one circuit into another by voltages developed across a *resistance* which forms a part of each of the two circuits. *See illustration*.

resistance drop.—The difference in voltage between the two ends of a resistance. Potential difference which is due to *ohmic resistance*.

resistance feedback.—A feedback of energy through a *resistance coupling*.

resistance loss.—*The I^2R loss*.

resistance variation method.—A method of measuring *high frequency resistance* by noting the change in current which is brought about when a known amount of resistance is added to a resonant circuit.

resistance wire.—A wire conductor having *ohmic resistance* high enough to allow its use in control of current and voltage.

resistive coupling.—*Resistance coupling*.

resistivity.—Specific resistance; the resistance between opposite faces of a unit cube of a substance, or the resistance of one milfoot of a conductor, these two being expressions of *volume resistivity*. Also expressed as *mass resistivity*, which see.

resistor.—A device containing *resistance* used in the operation, control or protection of electrical circuits.

RESOLUTION

resolution.—Same as *definition*.

resonance.—1—The condition in an *oscillatory circuit* or in any alternating current circuit having its (positive) inductive reactance equal to its (negative) capacitive reactance at a certain frequency called the *resonance frequency*. The two reactances balance and leave only the circuit *resistance* to oppose current flow, so that the current is maximum for a given applied voltage. The condition in which the current is in phase with the applied voltage. Circuits are "in resonance" when both or all show resonance at the same frequency. Compare *parallel resonance* and *series resonance*. 2—*Acoustic resonance*.

resonance characteristic.—A graph showing the relation between frequency and the current or voltage in an *oscillatory circuit*, covering frequencies on both sides of *resonance*. A graph showing changes of current in an oscillatory circuit when the frequency is varied from that of resonance. See illustration.

resonance circuit.—An *oscillatory circuit*.

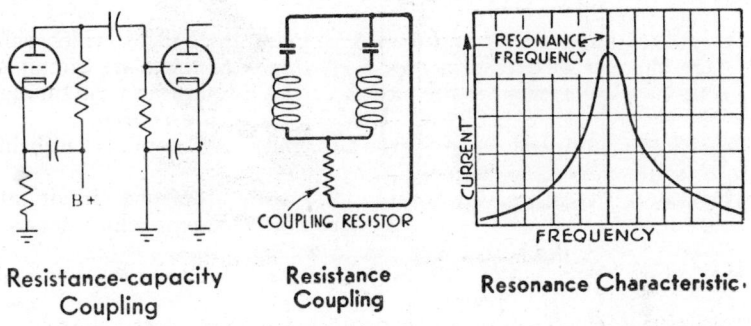

Resistance-capacity Coupling   Resistance Coupling   Resonance Characteristic.

resonance efficiency.—The ratio of the energy absorbed from *damped oscillations or waves* to the energy absorbed from equivalent *sine waves*.

resonance frequency.—The frequency at which *inductive reactance* and *capacitive reactance* balance or are equal in an *oscillatory circuit* or in any circuit containing inductance and capacity, the flow of current then being maximum for a given applied voltage of this frequency. The frequency at which current and applied voltage are in phase in a *reactive circuit*. The resonance frequency is equal to the reciprocal of 2π (pi) times the square root of the product of the circuit's inductance and capacity.

resonance indicator.—Any device which indicates by visible or audible means the condition of *resonance* at an applied frequency in an *oscillatory circuit*. Such indicators include small lamps, crystal detectors and various forms of alternating current voltmeters and ammeters.

resonance peak.—An increase of voltage upon application of a certain frequency to a device or circuit in which the combination of capacitance and inductance produce *resonance* at that frequency. *See illustration.*

resonance ratio.—The ratio of the current at *resonance* to the current in the same circuit with the capacity removed.

resonance transformer.—A transformer having a capacity connected to its secondary winding so that the secondary circuit shows *resonance* at the frequency of the voltage applied to the primary. Allows a secondary voltage higher than normal because of the elimination of reactive effects in the secondary.

resonant.—Exhibiting the condition of *resonance.* As generally used, the word has the same meaning as resonance.

resonant circuit.—An *oscillatory circuit* which exhibits *resonance* at the frequency existing in some associated part in an electrical or radio device.

resonant frequency.—A *resonance frequency.*

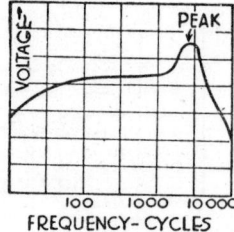
Resonance Peak

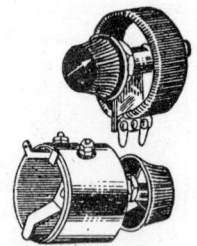
Rheostat

resonant line.—Two conductors positioned parallel to each other and supported continuously or at intervals by insulation. Inductance and capacitance in the line are resonant at frequencies corresponding to electrical lengths of a quarter-wave or a half-wave. For half-wavelength there is parallel resonance when ends far from the voltage source are open, series resonance with ends shorted on each other. For quarter-wavelength the open line is series resonant and the shorted line is parallel resonant.

resonating.—Assisting in the production of *resonance;* a part or portion of a resonant circuit or an *oscillatory circuit.*

resonator.—1—An enclosure of such size and form as to allow *acoustic resonance* at certain frequencies. 2—A name sometimes applied to an *oscillator.*

restoration.—D-c restoration.

retentivity.—The ability of a magnetic material to retain magnetism or to remain a magnet after the *magnetizing force* has been removed. The ratio of the *residual magnetism* to the maximum magnetism. Sometimes the word is used also in the meaning of *coercive force,* which see.

RETMA.—Abbreviation for Radio-Electronic-Television Manufacturers Association.

retrace.—Vertical retrace or horizontal retrace.

retrace blanking.—Prevention of vertical or horizontal retrace lines on a television picture tube or cathode-ray tube by application to cathode, first grid or second grid during vertical or horizontal retrace periods of a voltage pulse of such polarity and strength as prevents electron flow in the beam. Pulses are obtained from vertical or horizontal sweep or deflection circuits.

retrace period.—The *flyback period.*

reverberation.—Continuation of sound within an enclosure because

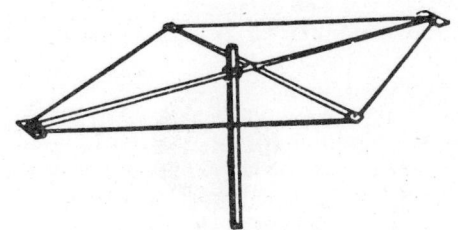

Rhombic Antenna

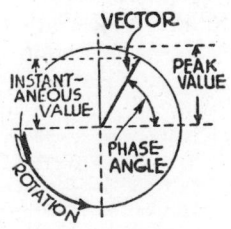

Rotating Vector

of *reflection* of the waves back and forth between surfaces forming the enclosure. The effect of numerous *echoes* following closely upon one another.

reverberation period, time or damping.—The number of seconds required for a sound to drop 60 *decibels* in intensity after the source of sound has ceased to act. See *Sabine's formula.*

reverse current relay.—A *relay* which opens a circuit upon reversal of direction of current flow.

R.F., r.f. or r-f.—Abbreviation for *radio frequency.*

R_f.—Symbol for *filament resistance* in ohms.

R_g.—Symbol for *grid resistance* from grid to cathode.

rheostat.—A device providing circuit resistance in a form allowing easy change of value. *See illustration.*

rhombic antenna.—An antenna having four conductors of equal lengths arranged in a parallelogram whose plane is horizontal. Signal takeoff is from a gap at the corner opposite the direction from which desired signals approach.

rhythm.—A recurring effect, such as successive beats in music.

ribbon antenna.—An *aerial* conductor which is of flat cross section.

ribbon microphone.—A *microphone* in which the member moved by sound waves is a light, thin aluminum ribbon located in a powerful magnetic field. Movement of this ribbon results in voltages from the device.

rim coil.—A field neutralizing coil.

ringing.—Any effect due to oscillation, intentional or otherwise. Ringing effects in television pictures may cause multiple images. A ringing coil in a sweep oscillator circuit provides a waveform allowing easier synchronization.

ripple filter.—A low pass filter designed to reduce the amplitude of *ripple voltage* currents, while passing direct current.

ripple percentage.—The ratio of the *ripple voltage* to the average value of the total voltage in a circuit.

ripple voltage.—A slight rise and fall of voltage in a circuit carrying direct current. The *alternating component* of the output voltage from a rectifier, filter or generator. The ratio of the effective value of the alternating component to the average value of the total voltage, usually expressed as a percentage.

R.M.A.—Abbreviation for Radio Manufacturers' Association.

r-m-s.—Abbreviation for *root-mean-square value*.

R_o.—Symbol for resistance of an *output circuit*.

rolling.—Continued upward or downward movement of television pictures, due to faulty vertical synchronization.

rolloff.—Descriptive of a frequency response having gradual increase of gain at the low-frequency end and gradual decrease at the high-frequency end, rather than having cutoffs or changes of gain which are sharp. A rolloff frequency response may have attenuations similar to those illustrated for low-pass and high-pass filters.

Rontgen rays.—The *X-rays*.

Rontgenography.—Photography by means of *X-rays*.

root-mean-square value.—The square root of the mean of the squares of all the instantaneous values in one cycle of alternating voltage or current. The *effective* value. With sine wave currents it is equal to the maximum value divided by 1.41421, or multiplied by 0.7071. Abbreviated *r-m-s*.

rotating vector.—A vector considered as rotating anticlockwise about one of its ends, the rate of rotation being equal to the cycles per second of an alternating frequency, each cycle being represented by one full turn of the vector. When the angle of the rotating vector with a horizontal line is equal to the *phase angle*, a projection of the vector on a vertical line represents the *instantaneous value* of the alternating quantity. See also *vector*. See *illustration*.

rotor.—A movable or moving part.

rotor plates.—The movable plates of a *tuning condenser*, generally the plates connected to ground or to the low voltage side of the tuned circuit. Compare *stator plates*.

R_p or r_p.—Symbol for *plate resistance*. In some cases the symbol R_p indicates external resistance in series with a tube plate, while the symbol r_p indicates the tube's internal plate resistance.

RTMA.—Abbreviation for Radio-Television Manufacturers Association. Now RETMA.

rumble.—Low, heavy, rolling sounds during phonograph reproduction, due to vibration in rotating parts of the apparatus.

R-Y signal.—The red-minus-luminance color-difference signal in color television. Equivalent to primary red minus the luminance or Y-signal. When combined with a plus Y-signal provides a primary red signal.

S

S.—Symbol for *photocell sensitivity*, also for *elastance*.

Sabine's formula.—A formula stating that the *reverberation period* in seconds is equal to one-twentieth of the room volume in cubic feet divided by the total *absorption units for sound* in the room.

saturation.—1—Plate current saturation. 2—Magnetic saturation. 3—A measure of the percentage of white mixed with a hue or color. High saturation refers to little or no white or to an intense color. Low saturation refers to little color and relatively much white. The degree of saturation does not alter the hue or predominant color wavelength. 4—The condition with which emitter current in point-contact transistor continues to increase while emitter voltage is decreasing.

saturation control.—In a color television receiver, an adjustment for amplitude of chrominance signal and thereby of color saturation.

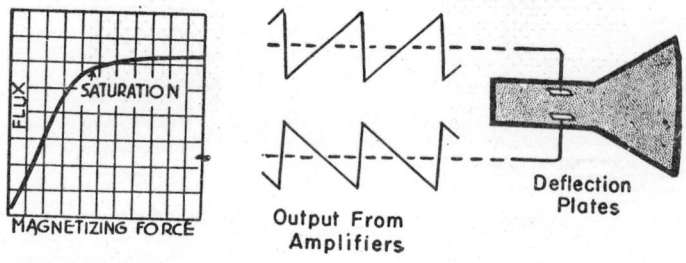

Saturation Curve   Sawtooth Wave

saturation current.—The maximum plate or anode current in a vacuum tube or photocell; the current which cannot be increased by further increase of applied voltage and which is limited by *electron emission* from the cathode. The current resulting when all the emitted electrons are drawn away from a surface, none falling back to it.

saturation curve.—A *magnetization curve* exhibiting the change in slope which indicates *magnetic saturation*. *See illustration.*

sawtooth wave.—A waveform of alternating voltage or current which changes rapidly from its peak of one polarity to the peak of opposite polarity, but relatively slowly during the reverse change. Employed for sweep or deflection of electron beams in television picture tubes and cathode-ray tubes.

SCANNING

scanning.—Conversion of lights and shadows of an image at the television transmitter into corresponding changes of voltage which are transmitted to the receiver and there used to reproduce lights and shadows on the picture tube which match those of the scene being viewed at the transmitter.

scratch filter.—A *low pass filter* used in connection with a phonograph pickup to attenuate the higher audio frequencies at which occur *needle scratch* resulting from motion of the pickup needle over the surface of a record.

screen.—A picture tube viewing screen. A screen grid.

screen grid.—The second grid in a pentode or in a screen grid tube, the grid immediately outside the control grid. In a pentode the screen grid is between the control grid and suppressor grid. The screen grid reduces capacitance between plate and control grid, and to this end is shorted to the cathode or to ground, so far as signal currents are concerned, by a large externally connected capacitance. To the screen grid is applied a direct potential highly positive with respect to the cathode, in order to draw electrons from the cathode through this grid and toward the plate.

Scanning

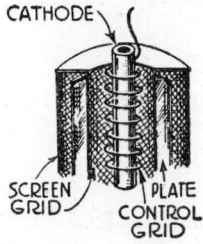
Screen Grid

screen grid tube.—A four-element *vacuum tube* having a cathode, plate and control grid, and in addition a perforated metallic shield almost completely surrounding the plate and called the screen grid. Through connections in the external circuits this screen grid is electrostatically grounded and maintained at such a potential that the capacity between plate and control grid is reduced to a small value. *See illustration.*

screen grid voltage.—The potential difference between a *screen grid* and the *cathode* of a vacuum tube. The symbol for the voltage at its source is E_d, for the voltage between screen and cathode it is E_s.

S-curve.—A frequency response showing variations of voltage with changes of frequency in the output of a frequency-modulation demodulator. A curve shaped somewhat like a letter S lying on its side.

SEARCH COIL

search coil.—An *exploring coil*.

second channel interference.—Interference from a signal which has a frequency differing from that of the desired tuned signal by twice the *intermediate frequency* of a superheterodyne receiver.

second channel selectivity.—The ability of a superheterodyne receiver to reject signals from a transmitter operating at a frequency separated from the desired tuned frequency by twice the *intermediate frequency*.

second detector.—In *superheterodyne reception*, a detector tube which follows the *intermediate frequency amplifier* and which produces audio frequency currents from the intermediate frequency voltages impressed upon its grid circuit.

second harmonic superheterodyne.—A type of superheterodyne receiver in which the second *harmonic* of the oscillator frequency is used in production of the *intermediate frequency*.

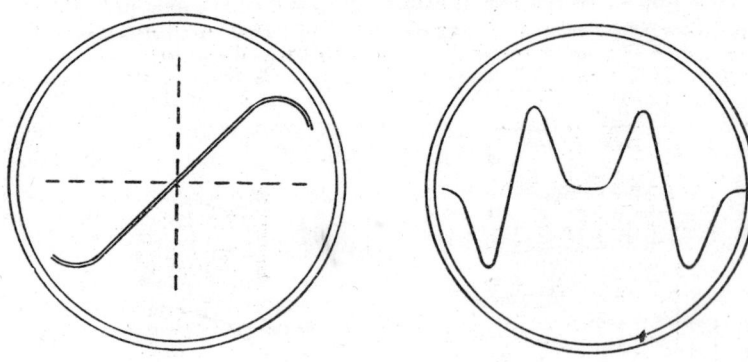

S-curve

secondary circuit.—A circuit which obtains its energy through coupling or connection with another (primary) circuit. A circuit connected to the output side of a transformer, a load circuit.

secondary emission.—Liberation of electrons from a cold body which is being struck or "bombarded" by other rapidly moving electrons which have been emitted from a cathode in the usual way. A form of *electron emission* which takes place from a vacuum tube plate or grid. For example, in a screen grid tube operated at a comparatively low plate voltage, secondary electrons liberated from the plate are attracted to the positively charged screen, thus lessening the total number of electrons remaining on the plate and correspondingly lessening the plate current.

secondary foci.—*Conjugate foci* which are at equal distances from the *optical center* of a lens.

secondary winding.—A winding connected in a *secondary circuit*.

SEEBECK EFFECT

Seebeck effect.—The effect producing *thermoelectricity*.
selectance.—A measure of selectivity. *Selectivity factor*.
selective consonance.—The quality of *resonance*, enabling a circuit to respond to some one frequency.
selective emission.—*Photoelectric emission* occurring in greatest amount only within a rather narrow range of light frequencies or wavelengths and falling off sharply on either side of this range. The only form of emission sufficient in amount to be practically useful in photoelectric work.
selective fading.—A kind of *fading* in which there is attenuation, during one interval, of the high frequencies and then, during another interval, attenuation of lower frequencies. The apparent effect is fading either of the high or else the low frequencies in the signal or the modulation.
selective network.—A *filter* circuit which attenuates certain frequencies.
selectivity.—The degree to which a receiver or a circuit differentiates between signals at different *carrier frequencies*. A measure of ability to respond to signals at one frequency while excluding those of all other frequencies, this ability being proportional to the change which occurs in the current of a *resonant circuit* when

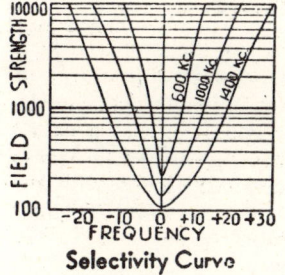
Selectivity Curve

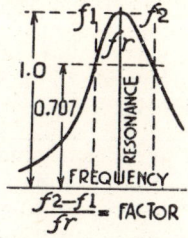
Selectivity Factor

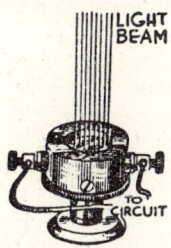
Selenium Cell

the capacity or the inductance is varied by a specified amount from the value at *resonance*. Selectivity increases with increase of the ratio of *inductive reactance* to *resistance* in a circuit, and it decreases as the *logarithmic decrement* becomes larger.
selectivity curve.—A curve showing the increase of *radio field intensity* which is required in order to maintain a *normal test output* from a receiver as the frequency of the applied signal is varied either side of the resonance point for a *standard test frequency*. The steeper the sides of the curve, the greater the receiver's selectivity. *See illustration.*
selectivity factor.—In a *resonant circuit*, the frequency of resonance divided into the difference between the two frequencies on either side of resonance at which the current has fallen to 0.707

times its peak value. Another selectivity factor is the reciprocal of the *logarithmic decrement*. *See illustration.*

selector.—The tuning control of a receiver.

selector switch.—A multiposition switch used for changing wavebands, tone values, instrument ranges or for connecting together any combinations of circuits.

selenium.—An element of which the electrical resistance is lowered when light strikes against it. A *photo-conductive* material.

selenium cell.—A device in which the element *selenium* is employed to change electrical resistance proportionately to the amount of light reaching the cell. Changes of current through the cell may be used for operation of relays or other electrical instruments. A *photo-conductive* cell. *See illustration.*

selenium rectifier.—A contact rectifier in which each element consists of a layer of selenium on an aluminum base with a conductive coating over the selenium. Electrons flow more easily from this coating to the selenium than in the opposite direction.

self-focus.—Automatic focus.

self-heated thermocouple.—A *thermocouple* which carries and is heated by current to be measured, instead of being indirectly heated through contact with or radiation from a separate conductor heated by the current.

self-heterodyne reception.—*Autodyne reception.*

self-inductance.—The property of a circuit by which it opposes any change in the rate of current flowing in it, or the property of a circuit in which a current changing its rate of flow produces in the circuit a voltage having a polarity opposite to that causing the original current flow. This voltage of self-inductance is called *counter-electromotive force.* Self-inductance is measured in *henrys* and is proportional to the lines of magnetic flux linked with the circuit per ampere of current causing the flux; 100,000,000 lines per ampere corresponding to a self-inductance of one henry. The symbol is L.

self-induction.—The action by which any change of current in a circuit (either an increase or decrease) produces an *electromotive force* which opposes the change of current.

self-oscillation.—*Self-sustained oscillation.*

self-shielded coil.—An *astatic coil.*

self-sustained oscillation.—Oscillations which are maintained by energy *feedback* from an output or plate circuit to an input or grid circuit.

Selsyn motors.—Two specially designed synchronous machines connected to a single A.C. source in such manner that any position, or any speed and direction of rotation of the rotor in one machine is accompanied by a similar position, or similar speed and direction of rotation of the rotor in the other machine.

semi-conductor.—A material having volume resistivity less than that of insulators but much greater than that of such good conductors as metals. Pure germanium would be classed as an insulator, but controlled addition of impurities makes it a semiconductor for transistors.

sensation level.—The number of *sensation units* or decibels by which a given sound power is above the power at the *threshold sound intensity* of audibility.

sensation unit.—The smallest change in power of a sound which can be distinguished by the ear as a change. A change of one *decibel* in the power level of a sound.

sense.—One particular direction of two opposite directions; one phase relation of two possible phase relations, etc.

sensitivity.—1—The response of a receiver to signals at the tuned frequency; the ability of a receiver to change *radio wave* signals into *sound* energy. Measured in various ways; as the ratio of the output power to the received power at various frequencies, or according to the input voltage required to produce a *normal test output,* or as the reciprocal of the *normal radio field intensity.*

sequential color television.—Any of several systems with which the primary colors, blue, green and red, appear one after another rather than at the same times from the phosphor screen or viewing screen, but in such rapid succession that colors blend to give visual impressions of all required hues or of white.

series aiding.—A connection of two *inductance coils* in series while they are in such positions that the fields of both act in the same polarity, the total field strength and total inductance being increased. Compare *series opposing*. See illustration.

series circuit.—A circuit with all its parts and conductors connected end to end so that all the current flowing through any one portion must flow also through every other portion. *See illustration.*

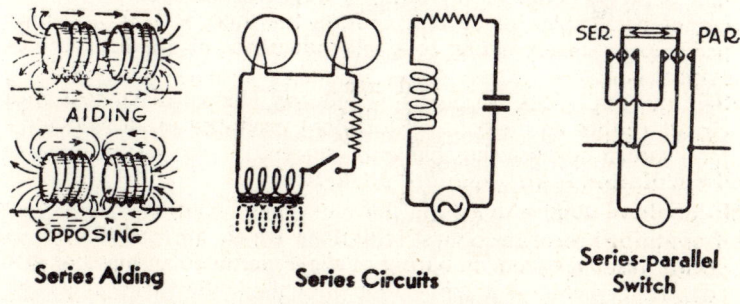

Series Aiding   Series Circuits   Series-parallel Switch

series compensation.—High-frequency compensation by means of a small inductor between the output of a detector or one amplifier and the input to a following amplifier, for the purpose of interposing inductive reactance between shunt capacitances in first and following circuits, thus reducing total effective capacitance.

series feed.—A connection of the direct current power supply for the plate of a tube to the end of an impedance which forms part of the active plate circuit, or part of the path for audio or radio frequency plate current. The direct current then flows to the plate through this impedance. Compare *parallel feed*.

series feed oscillator.—A vacuum tube oscillator employing *series feed* for the tube plate circuit.

series filaments or heaters.—A series connection of the filaments or heaters of two or more tubes, the voltage across the entire circuit being equal to the sum of the voltage drops across the filaments or heaters plus the drop in the intermediate connections.

series inductance of line.—The *self-inductance* possessed by the conductors forming a transmission line.

series loaded line.—A *loaded line* with the added reactances in series with the line.

series-multiple.—A number of parts connected in multiple or *parallel circuits*, these circuits then being connected together in a *series circuit*.

series opposing.—A connection of two *inductance coils* in series while they are in such relative positions that their fields act in opposite polarity, the total field strength being reduced and the *self-inductances* cancelling wholly or in part, leaving the *mutual inductance* as the chief characteristic. Compare *series aiding*.

series-parallel.—*Series-multiple*.

series-parallel switch.—A switch allowing two parts or two circuits to be connected together either in series or in parallel. *See illustration*.

series resistance.—1—A resistance connected in series with a circuit. 2—The *feed resistance* of a multiple tuned antenna. 3—An *equivalent resistance* (in series). 4—The *ohmic resistance* possessed by the conductors forming a transmission line.

series resonance.—The condition in a circuit which contains capacity and inductance in series, the values of the two being such that they produce *resonance* at the frequency of the applied voltage. The impedance to current flow is minimum, since the inductive reactance and capacitive reactance balance and leave only the circuit resistance. The current is in phase with the applied voltage. Compare *parallel resonance*.

serrated vertical pulse.—In the composite television signal, the vertical synchronizing pulse consisting of six parts separated by brief intervals called serrations, whose purpose is to allow continued horizontal synchronization throughout the entire pulse.

service area.—The localities in which the *radio field intensity* from a transmitting station is high enough to insure satisfactory reception of signals. Depends on the ratio of the signal strength to the interference field strength.

service band.—The *radio channels* employed for a certain kind or class of transmission.

set analyzer.—An instrument containing one or more meters and means for making necessary connections with receiver circuits, generally through tube sockets, so that voltages and currents in the several tube circuits may be read simultaneously or one after another.

shading coil.—1—A copper ring placed around the central pole of the field magnet in a *moving coil loud speaker* for the purpose of preventing hum pickup in the moving coil. 2—An extra winding used with alternating current electromagnets to prevent their pull from dropping to zero between alternations.

shading control.—A hue control.

shadow.—A region or area not reached or penetrated by light waves, sound waves or radio waves.

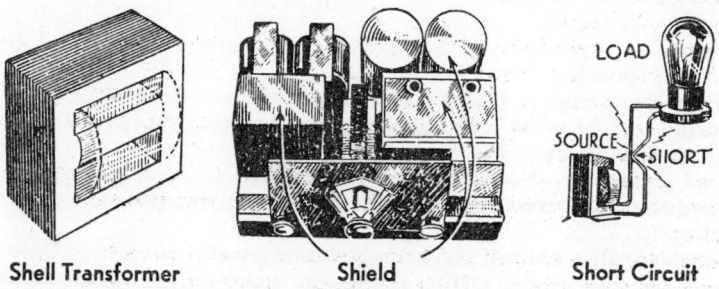

Shell Transformer   Shield   Short Circuit

shadow mask.—An aperture mask.

sharp cutoff.—Descriptive of a tube whose plate current is reduced to zero by making the control grid only a few volts negative with respect to the cathode, usually only four to eight volts.

sharp tuning.—The condition existing when a circuit or receiver has great *selectivity*.

sharp wave.—*Radiation* in which the frequencies are confined to a narrow band.

sharpness of resonance.—*Selectivity*.

shear vibration.—*Transverse vibration*.

shell transformer.—A transformer of which the magnetic *core* includes a straight central portion, surrounded by the windings, and extensions of the core which pass around the outside of the windings and join the ends of the straight central portion. *See illustration.*

shield.—A conducting metallic plate between electrical parts, or a complete metallic enclosure around such parts, this metal usually being grounded. External *electromagnetic* and *electrostatic fields* are prevented from reaching the enclosed or protected parts, and fields generated by the parts are not radiated through the metal barrier. *See illustration.*

shock excitation.—*Impulse excitation.*

short circuit.—A low resistance connection between the two sides of an electrical circuit, allowing current from a source to return to the source without flowing through the normal load. See *illustration.*

shorted out.—The condition in which a device or part of a circuit is made inactive by placing in parallel with it a conductor of very low resistance or impedance through which currents may flow around the part shorted out.

shorted line.—A resonant line whose conductors are conductively connected together at the end away from the voltage source.

shot effect.—A rapid fluctuation of *saturation current* in a vacuum tube, due to minute changes in the rate of electron emission. The result is tube noise in a receiver's audio output provided sufficient amplification follows the tube exhibiting the shot effect.

shroud.—A *shield.*

shunt.—1—One of the current paths in a *parallel circuit.* 2—A resistor connected in parallel with an ammeter to allow extension of the meter's current range. See *ammeter shunt.* Any similar resistance used to allow only a portion of the total current to flow through a device.

shunt capacity.—1—A capacity in parallel. 2—In a transmission line, the *distributed capacity* existing between the two sides of the circuit.

shunt circuit.—One of the branches in a *parallel circuit.*

shunt compensation.—High-frequency compensation by means of a peaker inductor in series with plate load resistance. The peaker becomes parallel resonant with stray and tube capacitances in the plate circuit to increase plate load impedance and tube gain at high frequencies.

shunt conductance of line.—The *conductance* existing between the two conductors or two sides forming a transmission line, and allowing a *leakage current* to flow.

shunt feed.—Connection of a load circuit to the plate of a tube through a capacitor which blocks direct current while passing signal currents, with direct current for the plate through a resistor or choke on the plate side of the blocking capacitor. Parallel feed.

side band or frequency.—A frequency equal to the sum of a carrier frequency and modulation frequency or to their difference. The upper (sum) and lower (difference) side frequencies or bands and the carrier may be transmitted. Otherwise the carrier may be suppressed, or else only one side band and the carrier may be transmitted.

side band transmission.—A system of transmission in which the *carrier current* is eliminated after modulation has taken place at the transmitter, the *side bands* being transmitted alone and received by apparatus which generates and supplies the missing carrier at the receiver. Carrier suppression. See *balanced modulator.*

SIDE FREQUENCY

side frequency.—A frequency which is produced by modulating the carrier current with a single additional frequency.
side wave.—A *side frequency*.
sign.—*Polarity*.
signal.—Any form of intelligence transmitted by radio waves or by

Signal Generator

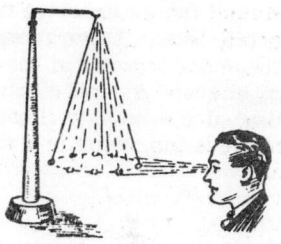
Simple Harmonic Motion

wire communication.
signal frequency.—The frequency of the *modulated wave* arriving at a radio receiver. Sometimes used as meaning the *modulation frequency*.
signal generator.—A device consisting of a *radio frequency oscillator* and an *audio frequency oscillator* arranged to furnish current at a known radio frequency, modulated, and to deliver a measured voltage only at the terminals of the generator without appreciable radiation at any other points. *See illustration*.
signal intensity.—The *amplitude* of a carrier current. The *radio field intensity*.
signal-to-noise ratio.—The ratio of desired signal strength to strength of random noise voltages which interfere with reception. Compare noise figure.
silent switch.—A switch, usually located at a distance from a receiver, and arranged to temporarily stop production of sound without cutting off the power supply to the receiver.
silver.—The metal having greatest conductivity. Sometimes used for switch contact points.
silver mica capacitor.—A fixed capacitor whose plates are coatings of silver on a mica dielectric.
silver screen.—A television picture tube viewing screen which is aluminized and to whose phosphors are added particles of silver.
simple harmonic motion.—The apparent motion of a point traveling at uniform speed around a circle when viewed from a position in the circle's plane, from the edge of the circle. *See illustration*. Approximately the motion of the lower end of a freely swinging pendulum.

simultaneous color television.—A system with which the primary colors, blue, green and red, appear at the same time on the picture tube phosphor screen or viewing screen. A three-gun color television picture tube is used in such a system. Compare sequential color television.

sine.—The sine of either of the smaller angles in a right triangle is the ratio of the length of the side opposite the angle to the length of the hypotenuse of the triangle. *See illustration.*

sine curve.—A curve representing rise and fall of the voltage induced in a conductor moving in a uniform magnetic field at constant speed around a circle whose plane is parallel to the direction of the magnetic lines. If measured from a horizontal line representing zero, the ordinate (vertical distance) of each point in

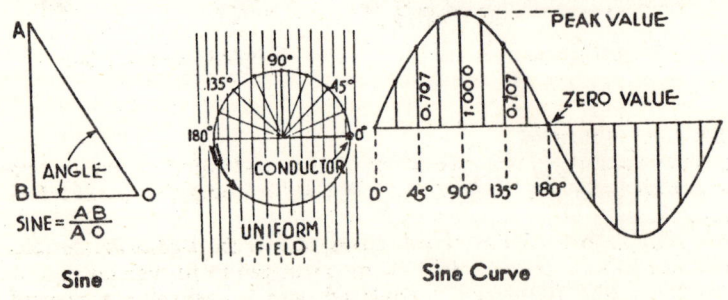

Sine   Sine Curve

the curve is equal to the *sine* of the angle through which the conductor has passed up to that instant, and the abscissa (horizontal distance) of this point is equal to the angle itself. *See illustration.*

sine galvanometer.—A *galvanometer* with which the *sine* of the angle of needle deflection is proportional to the current measured.

sine wave.—An alternating current or voltage in which the wave form is sinusoidal, varying according to a *sine curve*. This is the wave form assumed in all simple calculations of alternating current values.

single conversion.—Superheterodyne reception of the usual kind, with which modulated carrier frequencies are changed to intermediate frequencies having the same modulation by means of a single beating action in one mixer or converter. Compare double conversion.

single-gun color tube.—A color television picture tube having only one electron gun whose beam is directed successively to phosphors emitting the three primary colors. Employed in sequential color television.

single layer coil.—An *inductance coil* having its entire winding arranged in one layer with turns side by side.

SINGLE MAGNET ION TRAP

single magnet ion trap.—An ion trap so designed that ions and electrons are both turned from the tube axis by inclination or shape of the electron gun, with electrons brought to the beam path by the field of a single external magnet.

single-phase.—Descriptive of a circuit or apparatus in which there exists but one alternating voltage and its corresponding current.

single-pole switch.—A switch which opens and closes only one side of a circuit, having but one set of contacts. *See illustration.*

single side band transmission.—A transmitting system in which only one *side band* is radiated, the other band being suppressed. The *carrier wave* may be suppressed or radiated.

single silk covered wire.—Copper wire insulated with a single layer of silk threads. Abbreviated *S. S. C.*

single-throw switch.—A switch which may be closed in only one position, completing a circuit in only one path. *See illustration.*

single track recording.—Tape recording with only one magnetized sound track occupying all or part of the tape width.

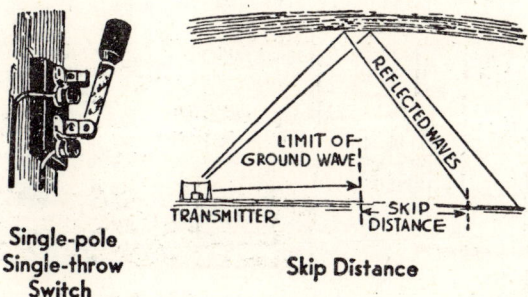

Single-pole Single-throw Switch

Skip Distance

sink.—A power consuming device or any device in which energy of one form is changed into energy of another form. A *transducer*.

sinusoidal.—Having the form of a *sine curve*.

skeleton form.—An open framework on which an *inductance coil* is wound.

skew.—Descriptive of a television picture whose rectangular outline is not parallel on sides, top, or bottom with that of a mask in front of the viewing screen.

skin effect.—An increase of *effective resistance* which accompanies increase of frequency in conductors carrying alternating current. *Eddy currents* and *counter-e.m.fs.* generated in the conductor's interior oppose flow of current in that portion, leaving the outer "skin" to carry most of the current, and thus reducing the current carrying ability of the conductor as a whole.

Skinderviken button.—A form of *microphone button*.

skip distance.—The space or region within which signals from a transmitter are not received. It extends from the farthest point reached by the *ground wave* to the nearest point at which the reflected wave or *sky wave* comes back to earth. The skip distance exists only with high frequency transmission and increases in extent with increase of frequency. *See illustration.*

sky wave.—The portion of a transmitter's radiated wave which is reflected back to the earth's surface from the layer of ionized gases near the top of the atmosphere. See *Heaviside layer.*

slab winding.—A winding placed on a thin, flat strip of insulating material. The inductance is low but the distributed capacity is rather high.

slide back voltmeter.—A *vacuum tube voltmeter* which measures *effective voltage*, this voltage being assumed as equal to the change required in *grid bias* voltage to return the plate current to the value it had before application of the measured voltage. The bias is altered by a *voltage divider. See illustration.*

slide contact.—A movable metal piece or brush which may be placed on various points in the length of a coil, a resistance or

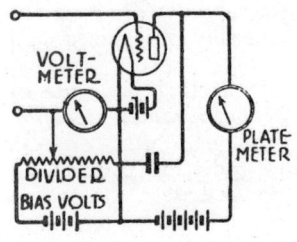

Slide Back Voltmeter

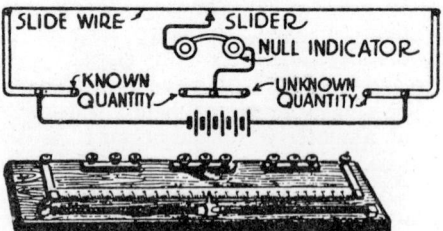

Slide Wire Bridge

other electrical element to include more or less of the element in an *active circuit.*

slide wire bridge.—A form of *Wheatstone bridge* in which the ratio is changed by a contact sliding on a single wire, the portions of this wire on either side of the slider forming the ratio arms of the bridge. *See illustration.*

slug.—A core of powdered iron which may be moved lengthwise inside the turns of an inductor to vary the effective inductance.

smearing.—Television picture distortion consisting of dark streaks on the right of bright areas or of light streaks thus following dark areas in pictures.

smooth line.—In telephony or telegraphy, a line having uniformly and continuously distributed electric elements; inductance, capacity, etc.

smoothing circuit.—A *low pass filter* following the rectifier in a *power unit.*

snow.—Television picture irregularities usually having the form of light or dark horizontal streaks of small size. The cause is excessive noise voltage picked up with received signals or produced in the receiver, or received signals are too weak in comparison with noise.

socket.—A *tube socket*.

socket adapter.—A device which allows a tube having one style of base to be used in a socket designed for a different base.

soft tube.—A tube within the bulb of which a small amount of gas remains, allowing slight *ionization* and increase of plate current with the tube in operation. Used as a *detector*. Compare *gassy tube*.

solenoid.—A spirally or helically wound conductor which exhibits the properties of a *magnet*.

sonometer.—An instrument for production of *sound waves;* a vibrating string stretched over movable bridges, the distance between which is measured by a fixed scale.

sonorous.—Having the ability to emit sound when struck. Having *acoustic resonance*.

S O S.—The letters which, in the radio telegraphic *code*, form the distress signal. Selected because of the easily recognized and distinctive sound of this signal.

sound.—A form of *wave motion* capable of affecting the sense of hearing; the waves being transmitted by *longitudinal vibration* of a material substance such as air, metal, glass, etc., whereas radiant energy, such as heat, light and radio, is transmitted through space without motion of any form of matter.

sound bars.—Alternate light and dark horizontal bands on television pictures, due to video signals being accompanied by any voltage at audio frequency.

sound carrier.—The frequency-modulated carrier which transmits television sound programs.

sound shadow.—A space within which sounds are not heard because of a screening object.

sound spectrum.—The frequencies included within the range of audible sound. *Audio frequencies*.

sound takeoff.—In a television receiver, the connection or coupling at which 4.5-megacycle frequency-modulated sound signals are obtained from some point following the video detector or from a separate sound demodulator.

sound wave.—The alternate compressions and rarefactions of the air or other medium through which sound is traveling. These waves cause motion longitudinally, or back and forth along the line of propagation. *See illustration*.

source.—A part which furnishes energy in any form; as electromotive force, light, sound, etc.

space charge.—Negative *electrons* which have been emitted from a cathode in a vacuum tube but which have not been drawn to the plate, remaining around the cathode as a *negative charge*. This negative space charge repels additional electrons emitted from the cathode. The space charge is made greater by a negative charge on a *control grid* or on any other electrode located near the cathode, and is made less by a positive charge on any such electrode.

space charge limited.—Operation of a vacuum tube with its cathode heated for normal electron emission and with element voltages low enough to allow a space charge to remain around the cathode at all times. The usual manner of operation.

space current.—The current between a vacuum tube anode or plate and the tube's cathode or filament, this current corresponding to the flow of *electrons* between these electrodes.

space factor.—1—In a winding, the ratio of the space occupied by the conductor with its insulation to the space occupied by the conductor alone. 2—In a magnetic *core*, the ratio of space occupied by iron to the total cubic content of the core.

space radio.—Radio transmission of signals through space by *radiation*, without the aid of metallic conductors to carry the signals. Compare *wired radio*.

space wave.—The *sky wave*.

spaced winding.—A coil winding in which adjacent turns are separated by an air space. *See illustration*.

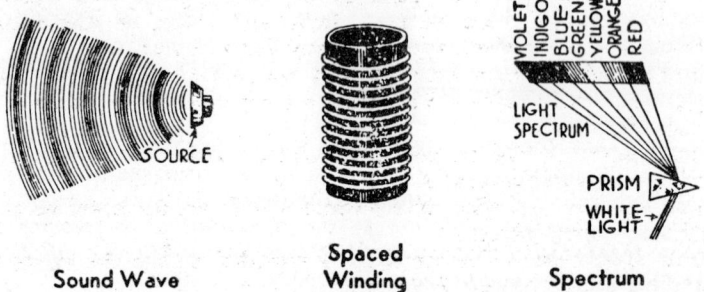

Sound Wave   Spaced Winding   Spectrum

spaghetti tubing.—Small diameter insulating tubing made of varnished cloth.

spark.—A momentary discharge of electricity across a space separating two electrodes. Compare *arc*.

sparking voltage.—1—The *potential difference* at which current will pass across an air gap. 2—The breakdown voltage in a *grid-glow tube*.

speaker.—A *loud speaker*.

speaker resonance. Excessively strong vibration of the cone and voice coil of a speaker within a limited range of audio frequencies, often near 100 cycles per second.

speaker transformer.—An iron core transformer which couples an audio output tube to a speaker voice coil and provides suitable impedance match.

specific gravity.—The weight of a substance expressed as the ratio of its weight to the weight of an equal volume of pure water. The weight of water is taken as unity, or as 1.0.

specific inductive capacity.—*Dielectric constant.*

specific value.—An electrical characteristic, such as resistance, conductance, etc., expressed in the quantity or value of such characteristic existing in a unit volume (centimeter cube) of the substance considered. The unit volume for conductors sometimes is the mil-foot.

spectrograph.—1—An illustration of a *spectrum*. 2—An instrument for forming a *spectrum*.

spectrometer.—An instrument which may be used for the indirect measurement of the *wavelength* of light.

spectrophotometer.—An instrument for measuring relative intensity of two like colors from different sources.

spectroscope.—An instrument which separates a light beam into a *spectrum*, into various colors and wavelengths.

spectrum.—An arrangement of rays of *radiant energy* in the order of frequency or wavelength. *See illustration.*

speech amplifier.—An *audio frequency amplifier.*

speech envelope.—An *envelope.*

speech modulation.—*Modulation* of a carrier by signals representing spoken words, music or other sounds. The modulation used for *radio telephony.*

spherical aberration.—An effect which results in failure of all the light rays passing through a lens to come to a single focus, the fault being due to curvature of the lens.

spherical face plate.—A face plate whose outside is everywhere curved like the surface of a sphere or ball of large radius, with no portions that are either plane or cylindrical.

spilling over.—The condition in the operation of a receiver at which *regeneration* changes suddenly into *self-sustained oscillation*, resulting in whistling in the audio frequency output.

split sound.—Dual channel sound.

spreader.—An insulating or supporting member holding two or more *aerial* conductors apart. *See illustration.*

spreading.—Movements of electrons and holes in a transistor crystal along irregular paths between terminals or junctions. The increased distances and times for travel cause maximum operating frequency to be lower than were the paths straight lines.

square law condenser.—A *straight line wavelength condenser.* A variable tuning condenser in which the capacity is proportional to the square of the wavelength to which the condenser's circuit is resonant.

SQUARE LAW DETECTOR

square law detector.—A tube detector with which variations of average plate current are proportional to the squares of the variations in signal voltage applied to the grid. A detector employing either *grid current detection* or *plate current detection* and operating at comparatively low plate voltages, as distinguished from a *power detector* or one providing *linear detection*.

square wave.—A voltage or current waveform with nearly instantaneous changes between maximum opposite amplitudes and with these amplitudes remaining nearly or quite constant between the changes. The square wave results from combination of a great many harmonics and sub-harmonics of the frequency at which the wave goes through its changes. For an amplifier or other device to pass a square wave without distortion requires that frequency response be excellent throughout a wide range, at least from one-tenth to ten times the frequency of the square wave.

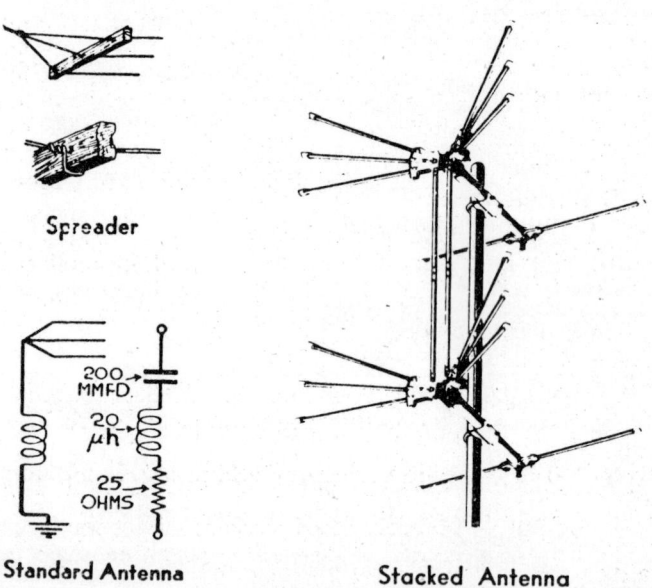

Spreader

Standard Antenna   Stacked Antenna

stability.—Freedom from regenerative feedbacks and resulting *self-sustained oscillation* in a radio frequency or audio frequency amplifier.

stabilizing.—*Balancing.*

stacked antenna.—An antenna having two or more bays mounted one above another with suitable phasing links between bays. The name sometimes is applied to bays mounted side by side.

stage of amplification.—An *amplifying tube* with its grid circuit and plate circuit.

staggered pairs.—Staggered tuning with alternate couplings tuned to the same frequency.

staggered tuning.—In a television intermediate-frequency amplifier system, tuning of alternate couplings at relatively higher and lower frequencies, all within a range such that overall response of the entire system has desired gains throughout the intermediate-frequency range and has a desired frequency response.

standard antenna.—A real or *artificial antenna* having definite electrical characteristics suitable for receiver testing. A series circuit with a resistance of 25 ohms, a capacitance of 200 microfarads and a self-inductance of 20 microhenrys. *See illustration.*

standard candle.—A form of candle which burns at a fixed rate and is used as a standard unit of *luminous intensity*. See *international candle.*

standard cell.—A *galvanic cell* which, under certain specified conditions, maintains a voltage constant enough to serve as a standard for comparison.

standard frequency signal.—A signal at a *standard test frequency.*

standard notation.—A system of writing large or small numbers as the product of two factors; one of which is a positive or negative power of 10, and the other a number with only one figure to the left of the decimal point. Thus 2.8×10^3 means to move the decimal point three places to the right, giving 2800.0; and 2.8×10^{-3} means to move the decimal point three places to the left, giving 0.0028.

standard ohm.—The *international ohm.*

standard test frequencies.—Frequencies used in receiver tests; these being 600, 800, 1000, 1200 and 1400 kilocycles, or sometimes only 600, 1000 and 1400 kilocycles.

standard test output.—The *normal test output.*

standard test voltage.—The *normal input voltage.*

standard wire gage.—The legal wire gage in Great Britain.

stand by.—To wait for further transmission of signals at a later time.

stand-by battery.—A *storage battery* held in reserve for emergency use.

stand-by circuit.—A broadly tuned receiving circuit which responds to any of several carrier frequencies on which messages may be received.

standing waves.—1—Points of minimum and maximum voltage and current which remain fixed in position along the length of a conductor from the terminus of which wave *reflection* takes place. The distance between two adjacent points of minimum voltage or current *(nodes)*, or the distance between two adjacent maximum points *(antinodes)* is equal to one-half wavelength. 2—A similar effect produced by interference of direct and reflected sound waves. *See illustration, page following.*

STAND-OFF INSULATOR

stand-off insulator.—An insulator of length sufficient to hold a conductor at a required distance from a building or other support. *See illustration.*

stat-.—A prefix used in the names of *electrostatic units.*

static.—1—*Interference* resulting from radio waves produced by atmospheric electrical disturbances, or from antenna currents produced by contact with the receiving aerial of electrically charged gases, rain, snow, etc. 2—A contraction of *electrostatic.*

static characteristic.—A curve showing the relations between steady values existing in a circuit or system. A curve made by application of direct currents and voltages to a system ordinarily operated with alternating or rapidly changing values, such as those of signal currents. Compare *dynamic characteristic. See illustration.*

static convergence.—Convergence of the electron beams in a three-gun color television picture tube to pass through openings at and near the center of the aperture mask. Convergence of undeflected beams.

static electricity.—*Electrostatic charges,* or electricity at rest.

static eliminator.—Any device intended to reduce the ratio of *static* to desired signal in the antenna circuit of a receiver.

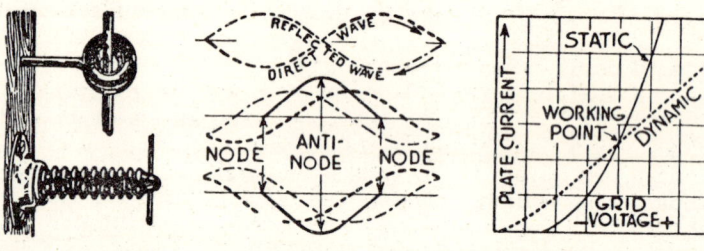

Stand-off Insulator   Standing Wave   Static Characteristic

static frequency transformer.—Two transformers with their primaries in series and their secondaries in series. Magnetizing force in one unit is strong enough to cause *magnetic saturation* of the iron during current peaks, thus producing in the secondary a distorted wave form which is combined with the wave from the other secondary to result in a new wave form of triple the frequency applied to the primaries.

static level.—The field strength of the combined effects of all forms of *interference* acting upon a receiving antenna and producing in the loud speaker output sounds which tend to overcome the desired signals.

stationary wave.—A *standing wave.*

stator.—The stationary portion of any device.

stator plates.—The fixed or stationary plates of a *tuning condenser*, generally the plates insulated from the framework and connected to the high voltage or grid side of the tuned circuit. Compare *rotor plates. See illustration.*

steady current.—A current which does not change, or one which changes regularly within certain limits or according to certain laws.

steady-state.—Descriptive of actions, such as currents or voltages, which continue without change or which repeat regularly after *transient phenomena* have ceased.

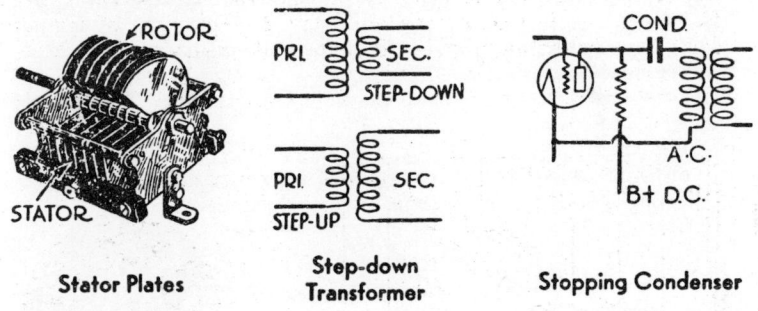

Stator Plates   Step-down Transformer   Stopping Condenser

step-down transformer.—A transformer in which the secondary voltage is less than the primary voltage, the secondary current then being greater than the primary current. *See illustration.*

step-up transformer.—A transformer in which the secondary voltage is greater than the primary voltage, the secondary current then being less than the primary current. *See illustration.*

stereophonic sound.—Binaural sound.

stopping condenser.—A condenser which impedes the flow of low frequency currents or prevents the flow of direct currents in parts of a circuit, while allowing comparatively free passage of high frequency currents. A blocking condenser. *See illustration.*

stopping potential.—The voltage to which the terminal potential of a photocell, a photo-glow tube or a grid-glow tube must be dropped in order to stop a *glow discharge* after such a discharge has once commenced. It is less than the *glow potential.*

storage battery.—Several *storage cells* electrically connected together.

storage cell.—A combination of electrodes and chemical electrolyte producing *electromotive force.* The chemical and electrical actions are reversible, the original chemical state being restored after discharge by forcing a charging current through the cell in a reverse direction. *Lead-acid* and *alkaline storage batteries* are in use.

straight line condenser.—One of several types of *tuning condenser* in which the amount, or the number of degrees, of rotation of the shaft is proportional to the change produced in *(a)* the condenser's capacity, *(b)* the wavelength to which the condenser's circuit becomes tuned, or *(c)* the frequency to which the condenser's circuit becomes tuned. These are called *(a)* straight line capacity condenser, *(b)* straight line wavelength condenser, and *(c)* straight line frequency condenser. *See illustration.*

strain.—A change in the electrical condition or in the shape or size of a body, the change being produced by an applied force.

strain insulator.—1—An insulator for mechanically connecting aerial guy wire sections, while electrically separating the sections into short electrical lengths to prevent their being *resonant* at frequencies near that of the antenna. 2—Any type of insulator capable of withstanding considerable pull or strain.

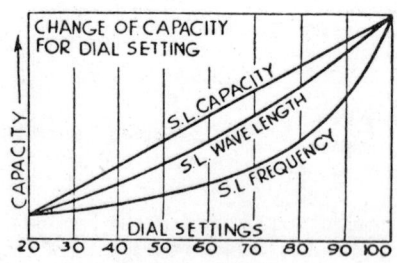

Straight Line Condenser

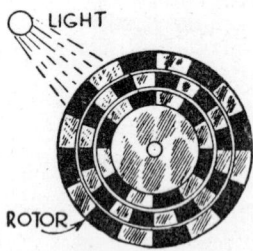

Stroboscope

strain pickup.—A phonograph pickup in which needle movement alters the cross sectional area of a conductor to vary its electrical resistance and cause corresponding variations of current and voltage.

stray capacities.—Capacities existing between parts of a circuit other than in its condensers. *Distributed capacities.*

stray field.—The portion of a magnetic field which spreads to some distance from a winding and which does not enclose the turns of the winding. *Leakage flux.*

strays.—*Static,* and also other electromagnetic field disturbances not caused by radio transmitters.

stress.—Force or forces which act to produce a change of shape or size in a body; also the body's opposition to such changes by counter-forces set up within it. The reaction which opposes electrostatic effects.

stroboscope.—An instrument for measuring *frequency* or studying any periodic action. The usual form consists of a rotating member illuminated by intermittent light. Regularly spaced lines or figures on the rotating member then appear stationary or appear to move in either direction, depending on the relation between speed of rotation and frequency of illumination. *See illustration.*

stub.—A short piece of transmission line providing inductive or capacitive reactance existing in such a line of length intermediate between those for resonance. Used for altering frequency responses. Also a resonant line used as an interference trap or else to increase gain of a connected circuit in a narrow range of frequencies.

stylus.—The pointed needle that cuts a groove on a phonograph disc during recording. A name sometimes applied to a needle which rests in the groove during reproduction.

subcarrier.—A voltage at frequency of 3.579545 megacycles per second generated at a color television transmitter, modulated with chrominance signals, then suppressed while chrominance side bands are transmitted. The subcarrier frequency is, however, transmitted in the form of bursts.

subcarrier oscillator.—A color oscillator.

sub-panel.—A flat surface used to support parts inside of a radio device.

substitution method.—A method of measuring an unknown value by first observing its effect in a circuit, then substituting in the circuit a similar but measurable value which may be adjusted to produce a like effect, the unknown value then being equal to the adjusted known value. In measuring *high frequency resistance,* current is measured in a resonant circuit containing the unknown resistance, the unknown is replaced with an adjustable resistance which is set to allow the same current, the two resistances then being equal.

subtractive process.—A method of making and projecting colored motion pictures in which shades and tones of desired colors are formed by filtering unwanted colors from white light.

sulphide rectifier.—A *contact rectifier* employing copper sulphide and either magnesium or aluminum as elements. Current flows easily from sulphide to magnesium or aluminum but with difficulty in the opposite direction.

super-audible.—A frequency higher than any audible frequency.

superheterodyne receiver.—A radio receiver employing the principles of *superheterodyne reception.*

superheterodyne.—A radio or television reception system with which modulated carriers of all channel frequencies are changed to a single intermediate frequency retaining the carrier modulation. This modulated intermediate frequency then may be strengthened by an amplifier which requires no changes of tuning regardless of the carrier frequency received. Modulated carriers are applied, with or without previous radio-frequency amplification, to a converter tube or mixer tube. To this tube is applied also a higher or lower frequency from an oscillator which is part of a converter or is in addition to a mixer. Carrier and oscillator frequencies beat together to produce the intermediate frequency at the output of mixer or converter. The oscillator circuit is variably tuned to the same number of kilocycles or megacycles above or below any received carrier frequency, thereby producing the same beat or intermediate frequency from all carriers.

super-high frequencies.—Defined as frequencies in the range from 3,000 to 30,000 megacycles per second.

super-regeneration.—A *regenerative detector circuit* in which maximum regeneration is used, but in which *sustained oscillation* is prevented by periodically applying to the tube's grid circuit a positive biasing voltage. This bias is generated in a separate oscillator circuit connected between the tube's plate and grid. *See illustration.*

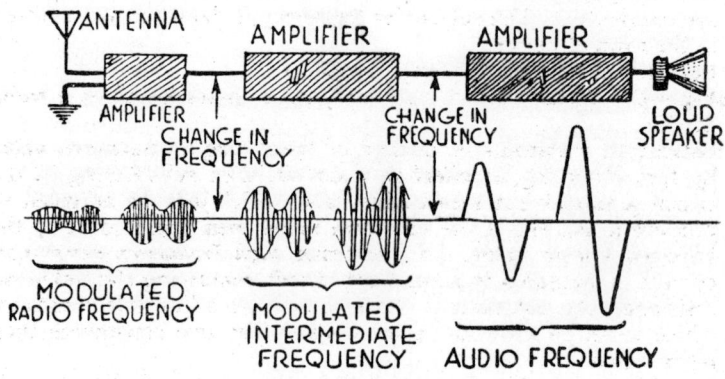

Superheterodyne Reception

supersonic.—Any wave motion or vibration at frequencies too high to cause audible sound.

suppressor grid.—The element of a pentode between its screen grid and plate, maintained at approximately cathode potential and serving to prevent secondary emission electrons from the plate reaching the screen grid while the screen grid is more positive than the plate.

surface barrier transistor.—A transistor having on opposite sides of an n-type base thin barrier layers of crystal in which electrons are neither surplus nor lacking, layers which are neither n-type or p-type. Metal electrodes on opposite sides of the crystal function as emitter and collector. Operation is somewhat similar to that of a p-n-p junction transistor.

surface leakage.—Escape of *current* over the surface of an insulator because of moisture, dust, etc., providing a conducting path on the surface.
surface resistance.—Resistance to flow of current over the surface of a material.
surface tension.—The *work function* of a metal.

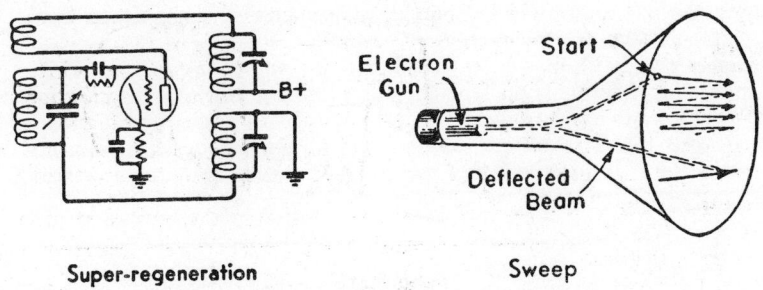

Super-regeneration   Sweep

surge.—A sudden increase of voltage or current in a circuit.
surge impedance.—*Image impedance.*
surge resistor.—A resistor which acts to lessen sudden rise of current which otherwise would occur upon application of voltage to a circuit of which the resistor is a part.
sweep.—Deflection of the electron beam horizontally or vertically in a television picture tube or cathode-ray tube. The word is used also to specify or describe any circuits or components that take part in the deflection process.
sweep generator.—A signal generator whose output voltage is frequency modulated, with frequency deviation through an adjustable range extending to a fraction of a megacycle or to many megacycles above and below a center frequency selected by a tuning control. Used in connection with an oscilloscope for observing frequency responses.
sweep oscillator.—In a television receiver or an oscilloscope, an oscillator producing sawtooth voltages which, after amplification, cause electrostatic deflection or produce sawtooth currents for magnetic deflection.
swing.—Grid swing.
swinging.—1—A frequency variation due to momentary changes in a transmitter's circuits. The variation may be due to motion of the aerial with reference to the ground, causing a change in antenna circuit capacity. 2—*Hunting* of synchronous alternating current machines.
switch.—A device for closing, opening or changing the connections in an electric circuit.
sympathetic vibration.—Periodic vibration at a body's natural

SYNC

frequency, the motion being produced by waves from another body which is vibrating at that frequency.

sync.—An abbreviation for synchronizing. The word is used to specify or describe any circuits or components which take part in timing or synchronizing the deflection of electron beams in television picture tubes to correspond with periods during which picture signals are received.

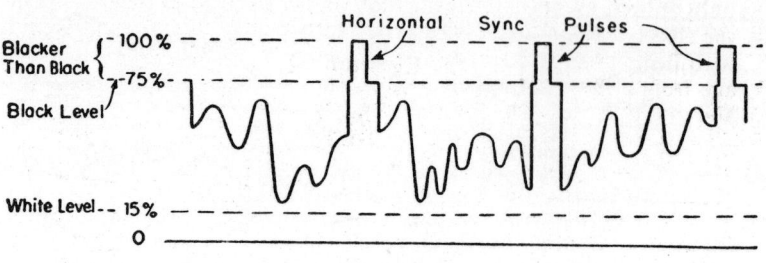

Sync Pulse

sync clipper.—A name sometimes applied to a sync separator, also to a sync limiter.

sync inverter.—An inverter tube employed to obtain from its output sync pulses of opposite phase or reversed in polarity with respect to pulses at its input.

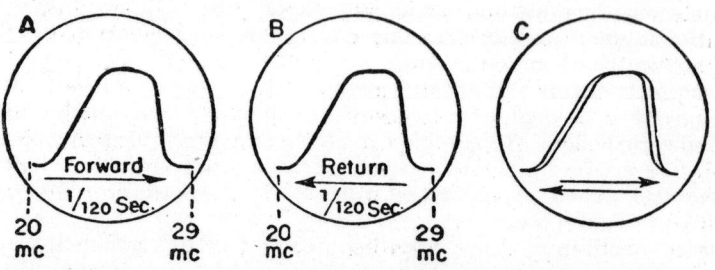

Synchronized Sweep

sync limiter.—A tube operated with element voltages such as cause either plate current cutoff, plate current saturation, or both, for the purpose of preventing sync pulses applied to its input from exceeding certain amplitudes in the output.

sync pulse.—In the composite television signal, various voltage variations of approximately square waveform, whose purpose is to start each cycle of operation in vertical and horizontal sweep oscillators.

sync separator.—A tube and associated circuits from whose output a complete composite television signal or else the portion carrying sync pulses is applied to circuits leading eventually to the sweep oscillators.

synchronize.—To bring electrical actions, voltages, currents, or their phases into such time relation with one another that desired performance or results are obtained, or an in-phase condition is obtained.

synchronized sweep.—A means for timing horizontal deflections of the electron beam in an oscilloscope cathode-ray tube to correspond with beginnings and endings of frequency deviation cycles in voltage from a sweep generator. When such timing is correct, forward and return traces on the cathode-ray tube are superimposed to appear as a single frequency response curve.

synchronous.—The relation between actions which are taking place at the same time, and usually in the same manner.

synchronous machine.—An alternating current machine in which the frequency is proportional to the speed of rotation.

synchronous motor.—A *synchronous machine* acting as an alternating current motor, the speed of rotation keeping step with the alternating current supplied to it.

synchronous detector.—A chrominance signal demodulator employed in some color television receivers. One detector tube or tube section recovers each of the phases in a chrominance signal or recovers either the I- or the Q-signal.

synthetic bass.—The impression, in the ears of listeners, of hearing bass notes or low audio frequencies not actually present in the reproduction, but some of whose harmonic frequencies are present. An amplifier having poor gain at low frequencies but having distortion which produces fairly strong harmonics of low frequencies may provide a synthetic bass effect.

syntony.—The condition under which two *oscillatory circuits* have the same natural frequency, the two circuits then being in syntony.

T

tandem condenser.—Two or more *tuning condensers* placed end to end, operated by a single rotor shaft passing through all units or by connecting the separate rotor shafts together. The stator plates of each unit are electrically independent of the other stators.

tangent galvanometer.—A direct current measuring instrument consisting of a magnetic needle mounted within a coil winding through which flows the measured current. The current is very nearly proportional to the tangent of the angle to which the needle is deflected by the field of the coil. *See illustration.*

tank circuit.—A name for the resonant circuit of an oscillator, and sometimes for a tuned circuit in an amplifier unit. So called because energy is stored in the inductance and capacitance, as in a tank.

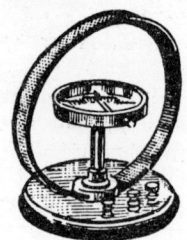
Tangent Galvanometer

T-antenna

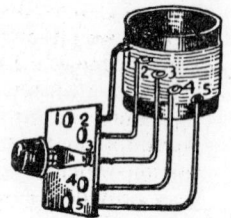
Tap Switch

tantalum rectifier.—An *electrolytic rectifier* employing the metal tantalum for the active electrode and a sulphuric acid solution for the electrolyte.

T-antenna.—An antenna consisting of horizontal aerial conductors to which connection is made at the approximate center, forming a T-shaped structure. *See illustration.*

tap.—A connection to one of the turns in an inductance coil, to a midpoint in a resistor or to any point between the ends of a circuit.

tap switch.—A multi-point switch that places additional turns in the *active circuit* of an inductance coil or other electrical element. *See illustration.*

tape recorder.—A device for magnetizing areas on a recording tape in patterns corresponding to amplitudes and frequencies of recorded sound. When the tape is used for reproduction with an audio amplifier the magnetization induces sound voltages and currents.

tape speed.—The rate at which recording tape is moved during recording or reproduction, the rate being constant and usually between two and 30 inches per second.

taper plate condenser.—A variable *tuning condenser* in which the entering edges of the plates are thinner than the trailing edges, allowing construction of a *straight line condenser* of wavelength or frequency type within limited space. *See illustration.*

tapered potentiometer.—A continuously adjustable potentiometer in which resistance varies at a non-uniform rate along the element, the change being greater or less for equal slider movement at various points. Various tapers or rates of resistance change are available to suit special requirements of control circuits in receivers and instruments.

tape-wound resistor.—A form of conductor used for resistors, formed by weaving a tape with metal wire for the woof and fabric

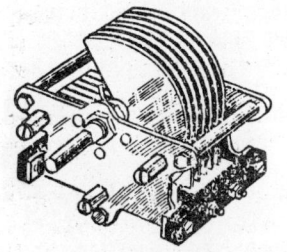
Taper Plate Condenser

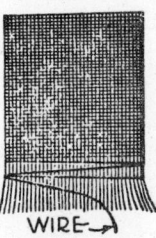
Tape-wound Resistor

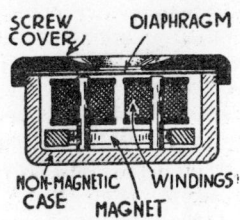
Telephone Receiver

threads for the warp. *See illustration.*

tapped winding.—A coil winding with connections brought out from turns at various points so that more or less of the winding may be made active.

tau (τ).—Greek letter symbol for *time constant*.

tear out.—Sharp horizontal displacement of portions of television pictures at top or bottom.

telegraph.—A communication system including apparatus for production and transmission of signals through electric circuits by means of changes in current, and associated apparatus for reproduction of the signals in audible or visible form at the far end of the circuit.

telegraph modulated wave.—A *continuous wave* varied in frequency or in amplitude by *keying*.

telegraph repeater.—Apparatus for repeating signals from one telegraph circuit into another.

telegraph sounder.—An instrument operated by electromagnets to produce telegraph signals in sound.

telegraphone.—A sound recording and reproducing device in which a moving wire, tape or disc of steel is unevenly magnetized at different points in its length or surface by means of signal currents. This unevenly distributed magnetism then is used to produce inductive effects which make sounds in a reproducer.

telegraphy.—The method of signalling which employs various arrangements of short and long, dot and dash, impulses sent by electrical means such as conducting wires, space radio and the like.

telemeter.—A device which furnishes visible or audible indications of the operation and functioning of apparatus located at a distance.

telephone.—A communication system which includes apparatus for the variation of electric current in accordance with sound waves, also a circuit carrying the current and associated apparatus for the reproduction of sound from the variations in current.

telephone condenser.—1—Any condenser made with paper dielectric and metal foil plates, rolled and wax impregnated. 2—A *fixed condenser* connected across a *telephone receiver* to bypass high frequency currents. Such a condenser may be charged by a wave train and then discharge through the receiver to produce one audible pulse.

telephone cord.—A very flexible double conductor, usually of tinsel or of small braided wires, covered with fabric and suitable for carrying currents to *telephone receivers*.

telephone receiver.—A device which changes variations of electric current into sound waves equivalent in form to the current waves. An *electro-acoustic transducer* driven by an electric system and supplying power to an acoustic system. A *loud speaker* or a *headphone*. See illustration.

telephone repeater.—An *audio frequency amplifier* placed in a long transmission line, with suitable input and output impedance matching transformers, so that the signal strength is increased before passing farther in the line.

telephony.—A method of reproducing sounds, especially those of the voice, at a distance. See *telephone*.

telephotography.—Transmission of still pictures by radio or wire circuits. *Facsimile transmission* or *picture transmission*.

television.—The transmission and reproduction at a distance of images of moving objects, the transmission means being *space radio*, *wired radio* or other electric circuits. Transmission of a succession of images in such manner as to produce the illusion of continuous motion.

telharmonium.—A musical instrument from which desired effects are produced by the mixing of various *fundamental tones* and *overtones* in circuits which include special types of transformers and impedance control devices. The original tones are secured from inductor alternator machines working at the required frequencies.

tell-tale.—A signal, lamp, or equivalent device which indicates the continued operation or the failure of electrical apparatus or circuits.

temperature coefficient of resistance.—The change in a material's *resistance* per centigrade degree of rise in its temperature, the change being expressed as a fraction of the resistance of the material at a reference standard of temperature. The reference temperature usually is either 0° or 20° centigrade.

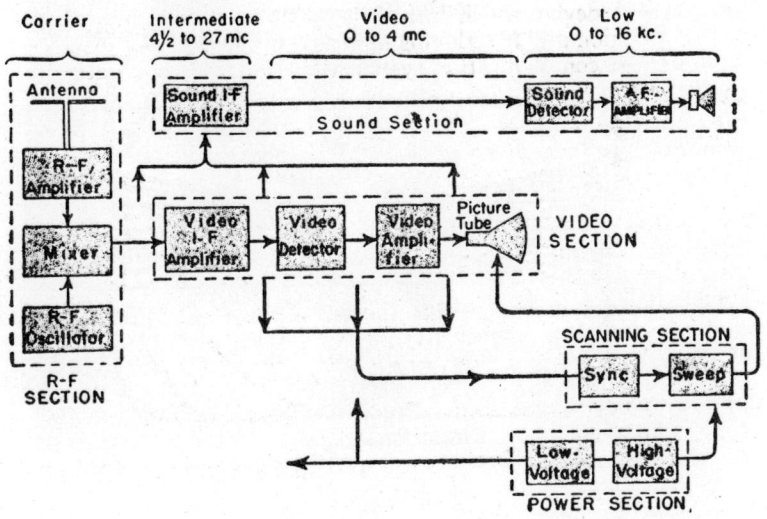

Television Receiver

temperature compensating capacitor.—A capacitor whose dielectric is of such material that capacitance either decreases or increases at a known rate with rise of temperature. The variation is specified in micromicrofarads per microfarad of nominal capacitance, per centigrade degree of temperature rise. The coefficient is negative for decrease of capacitance or positive for increase of capacitance with rise of temperature. Used chiefly to prevent frequency drift.

temperature limited.—Operation of a vacuum tube with cathode temperature so low or voltages on other elements so high that electrons are drawn away from the cathode as rapidly as emitted, leaving no space charge.

temperature refraction.—Sound wave *refraction* due to movement of sound waves through regions at different temperatures.

temperature relay.—A relay which operates because of a change in temperature.

tension.—*Electromotive force* or *potential difference*.

terminal.—A screw, clamp, lug or other suitable means for attaching conductors together or to circuits.

terminal impedance.—The *impedance* of a circuit or electrical device as measured between the input terminals or between the output terminals.

terminal voltage.—The voltage difference between the terminals of an electrical source or load.

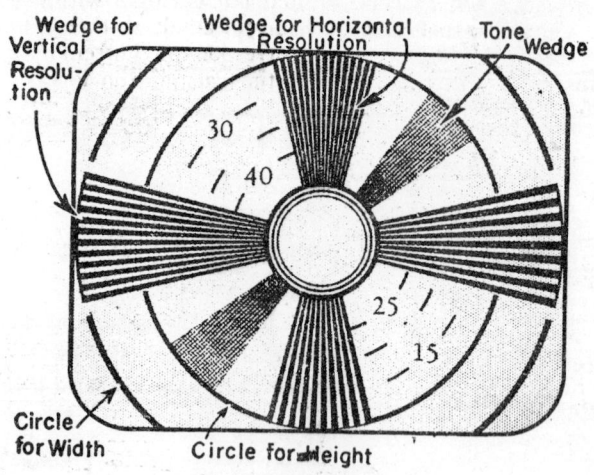

Test Pattern

terminated cable.—An instrument cable fitted with a matching pad.

test pattern.—A stationary design or geometrical pattern, chiefly of lines, circles and bars, transmitted by a television station for convenience in testing and adjusting receivers.

tetrode.—A tube or transistor having four elements which take part in admission, emission, control, and collection of electrons in tubes and transistors, and of holes in transistors.

tetrode transistor.—A point-contact transistor on whose crystal rest three metallic contacts. The name is applied also to a junction transistor having two terminal leads for the base element.

thermal noise.—Noise effects caused by amplification of small voltages which result from random movement of free electrons in any conductor, most noticeably in resistors. Movement and noise increase with rise of temperature.

theatre amplifier.—A *public address system* with its reproducers, designed for use in theatres and halls.

THEREMIN

Theremin.—A form of *musical oscillator*.

thermal ammeter.—1—A current measuring instrument utilizing the heating effect of a current to produce expansion in a *hot wire meter* or a thermoelectric effect in a *thermocouple instrument*.

thermionic.—Relating to *electron emission* under the influence of heat.

thermionic current.—The electric current passing between a cathode and a positively charged electrode in a *vacuum tube*.

thermionic emission.—*Electron emission* due to the action of heat.

thermionic oscillator.—A *vacuum tube oscillator*.

thermionic rectifier.—A *rectifier* having one or two cold plates or anodes and a hot cathode or filament enclosed within a vacuum bulb. Current can flow only from an anode to the cathode. The entire electron stream and the corresponding current flow are due to passage of electrons through the vacuum, no ionization being employed to increase the current. *See illustration*.

thermionic tube.—A vacuum tube in which *electron emission* takes place from a heated cathode.

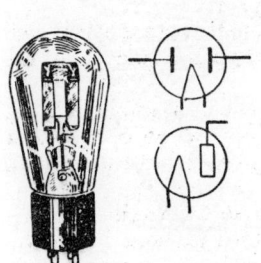
Thermionic Rectifier

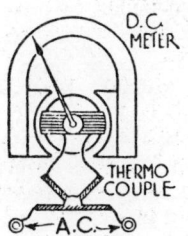
Thermocouple Instrument

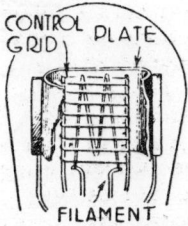
Three-element Tube

thermistor.—A resistor having large negative temperature coefficient.

thermoammeter.—A *thermocouple instrument*.

thermocouple.—A junction between two dissimilar materials which, when heated, produces an *electromotive force* across the junction. With the opposite extremities of the materials also joined, there is a flow of current when the two junctions are maintained at different temperatures.

thermocouple instrument.—A current measuring instrument in which a *moving coil instrument* is actuated by current from a *thermocouple* heated either by the measured current flowing through it, by contact with a conductor heated by the measured current, or by being in close proximity to such a heated conductor. *See illustration*.

thermoelectric current.—Current produced by action of a *thermocouple*.

thermoelectric power.—The ability of a metal, when used in a *thermocouple*, to produce electromotive force. The voltage produced by a temperature difference of one degree centigrade in a thermocouple using the metal as one part and lead as the other.

thermoelectricity.—Electricity produced by the direct action of heat, as in a *thermocouple*.

thermoelectromotive force.—1—An electromotive force produced by heating the junction between two different materials. 2—*Thermoelectric power*.

thermoelement.—A *thermocouple* and a heating member arranged for current measurement.

thermogalvanometer.—A current measuring instrument including a *moving coil instrument* connected to a *thermocouple* which is heated by a conductor carrying the current.

thermojunction.—A *thermocouple*.

thermostat.—A device which, upon cnange in the degree of heat applied to it, exerts mechanical force used to operate a switch, a valve, or other control.

theta (θ).—Greek letter symbol for *phase angle difference*.

thirty-degree cut.—A *Y-cut* for a quartz crystal.

Thomson effect.—The carrying of heat or cold by an electric current between parts of iron or copper pieces when the parts are at different temperatures.

thoriated filament.—A vacuum tube filament containing a compound of *thorium* which forms a surface layer one atom deep and allows a greater *electron emission* with a given temperature than can be had from a plain tungsten filament.

three-element tube.—A tube containing three electrodes; a filament or cathode, a plate or anode, and a third element called the control grid or simply the grid. The potential of the grid with reference to the cathode controls the electron flow and current flow between plate and cathode. *See illustration*.

three-gun color tube.—A color television picture tube in which are three individually controlled electron guns, one each for excitation of phosphor dots of the three primary colors, blue, green, and red. Used in a system of simultaneous color television.

three-phase.—Descriptive of an *alternating current* system in which are three separate e.m.fs. and three currents at one time, the angles between phases being 120 degrees.

three-way switch.—A switch which connects one conductor to any one of three other conductors.

three-wire system.—A circuit including three conductors between the outer two of which there is a voltage twice as great as between either of the outside wires and the center wire.

threshold current.—The current flowing in a *gria-glow* tube at the instant of *breakdown voltage;* the maximum current with which there still is a stable relation between applied voltage and flow of current.

threshold sound intensity.—1—Threshold of audibility is the sound intensity which is just sufficient to affect the sense of hearing. It varies with frequency of the sound waves. 2—Threshold of feeling is the intensity of a sound wave at which it commences to affect the sense of feeling; the upper limit of hearing with respect to intensity. See *auditory sensation area*.

thyratron.—A three-electrode *gaseous tube* containing cathode, anode and grid; a current conducting arc or glow discharge being controlled by the grid. The tube is used as a rectifier (changing A.C. to D.C.), as an inverter (changing D.C. to A.C.) and as a current or power control device, usually operating according to phase lag of grid voltage behind anode voltage.

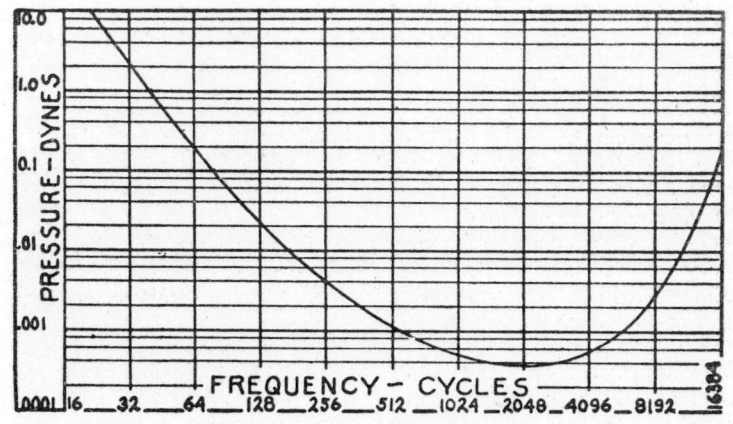

Threshold of Audibility

Thyrite.—A conductive material whose resistance decreases with increase of applied voltage. Doubling of voltage causes resistance to drop in the ratio of 12.6 to 1, throughout a total possible ratio of about ten million to one.

tickler coil.—A coil or winding in which flows current from the output side of an amplifier, detector, or other circuit, and which is coupled to the input for the purpose of providing regeneration or oscillation.

tight coupling.—Close coupling.

tilt control.—An adjustment for varying the waveform of parabolic current or voltage in a color television receiver, shifting the current or voltage peak one way or the other.

tilted gun ion trap.—An ion trap with which initial direction of both ions and electrons is determined by having the electron gun axis at an angle to the center axis of the picture tube, and with

TIMBRE

which electrons are brought into the beam path by the field of a single externally mounted permanent magnet.

timbre—The combination of *overtones* which distinguishes one voice from others, or one musical instrument from others when both produce the same *fundamental tone*.

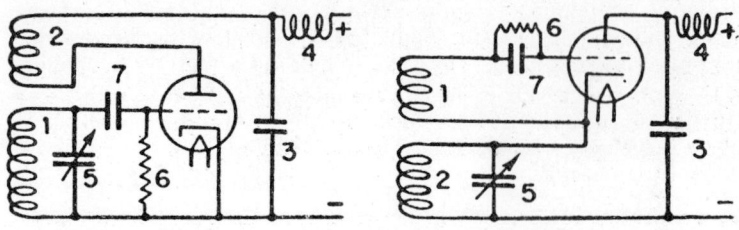

Ticklers

time constant.—1—In a *capacitive circuit*, the number of seconds required for the capacity to receive 63.2 per cent of its full charge after the e.m.f. is applied. With steady applied voltage the time constant is equal to the product of the circuit's *capacity* in farads and its *resistance* in ohms. 2—In an *inductive circuit*, the number of seconds required for the current to reach 63.2 per cent of its final value after the e.m.f. is applied. With steady applied

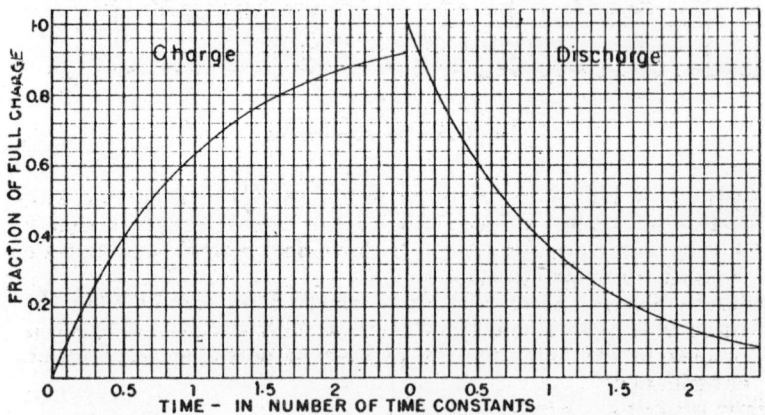

Time Constant

voltage the time constant is equal to the circuit's *inductance* in henrys divided by its *resistance* in ohms. The symbol for either time constant is the Greek letter tau (τ).

TIP JACK

tip jack.—A small receptacle into which fits a metal tipped conductor to complete a circuit. *See illustration.*

toggle switch.—A switch operated by moving a small projecting lever. *See illustration.*

tolerance.—A percentage by which actual value of a resistor, capacitor, or other electrical unit may differ from the nominal or rated value, either plus or minus.

tolerance frequency.—An additional range of frequencies above and below those normally used by a modulated carrier, this range caring for unavoidable variations from the assigned *carrier frequency.*

tone.—The character of a sound, especially with reference to regularity of wave form, harmonics, and other attributes affecting the pleasantness of the sound to the hearing.

tone chamber.—The air space enclosed by the walls of a horn or sound projector; also the *projector* itself. *See illustration.*

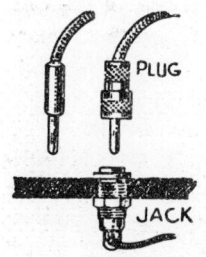
Tip Jack

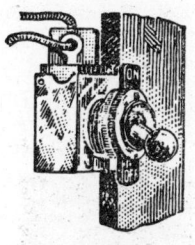
Toggle Switch

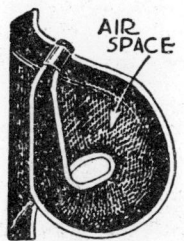
Tone Chamber

tone color.—*Timbre.*

tone control.—In an *audio frequency amplifier,* a method of emphasizing either low notes or high notes at will. Various methods are employed; such as bypassing higher frequencies to ground through an adjustable capacity, the use of tuned filter circuits for reduction of either high or low frequencies, the control of energy in transformers having desired frequency response, etc.

toroid.—The surface of a ring which has a circular cross section.

toroidal coil.—A *closed field coil* in which the winding as a whole is formed into a nearly closed circle, the ends of the winding being brought together to form the coil into a large ring. *See illustration.*

torque of instrument.—The turning force produced in an instrument's movement by the current or voltage to be measured; also the force developed in the mechanism tending to return the instrument pointer to zero position.

torsional vibrations.—Motion of disturbed particles in a body or substance as it is twisted around its axis.

total emission.—The current resulting from drawing away from a cathode of all the emitted electrons. See *saturation current*.

trace.—A visible line on the viewing screen of a television picture tube or cathode-ray tube caused by travel of the electron beam over the internal phosphor surface.

tracer.—Threads of contrasting color woven into the principal color of wire insulation for identification of a conductor with its circuit.

tracking.—Adjustment of two or more resonant circuits so that all are simultaneously tuned to required operating frequencies by a single control. Such an adjustment for the oscillator in a superheterodyne tuner.

trailers.—Smears in television pictures.

transconductance.—The ratio of change of current in the circuit of one vacuum tube electrode to the change of voltage on another electrode in the same tube, the voltage change being the cause of the current change, and other potentials remaining constant. The term has the same meaning as *mutual conductance* and is used when referring to *four-* and *five-element tubes*.

transducer.—Any device taking power from one electrical, acoustical or mechanical system and furnishing power to another system of the same or different type.

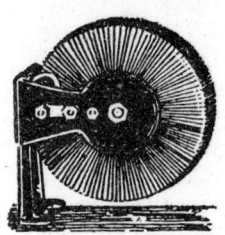

Toroidal Coil

Transformer

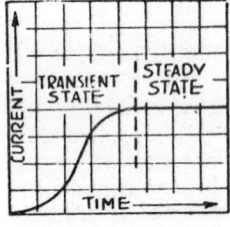

Transient Current

transfer characteristic.—A line or curve on a graph showing the effect on voltage or current in one element of changes of voltage or current at another element in a tube or transistor when the change is of voltage at either element and of current at the other.

transfer voltage.—In a *grid-glow tube*, the value of applied voltage at which there is a transfer of the discharge from the cathode-grid path to the cathode-anode path.

transformation ratio.—The ratio of the primary to the secondary voltage of a *transformer*.

transformer.—Two windings with a common *magnetic circuit*, allowing an alternating or fluctuating current in one winding to produce an alternating electromotive force in the other winding

TRANSFORMER COUPLED AMPLIFIER

because of *mutual inductance*. Energy transfer between the two electric circuits results from the *mutual induction* in the transformer. *See illustration.*

transformer coupled amplifier.—An amplifier using inductive coupling by means of a *transformer* between the plate circuit of one tube and the control grid circuit of the following tube.

transformer coupling.—*Inductive coupling.*

transformer loss.—The *transmission loss* which would be eliminated at a junction between circuits or networks by the use at that junction of an *ideal transformer* to match the impedances.

transformer oil.—A petroleum or mineral oil used for filling transformer cases to provide insulation and cooling.

transformer ratio.—Generally the *turns ratio*. Sometimes the *voltage ratio*.

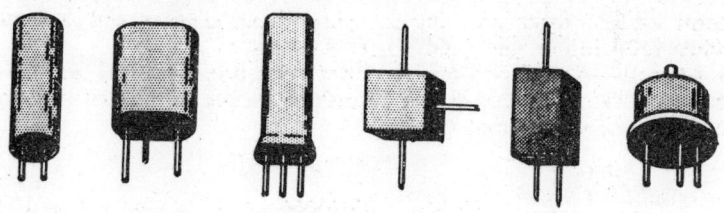

Transistors

transformerless receiver.—A receiver in which a.c. power line voltage is applied directly to tube heaters or filaments in series with one another, also to a power rectifier and filter circuit without first being stepped up or down by a transformer.

transient current.—A current which changes in an irregular manner, in which similar changes do not occur in any regular order. *See illustration.*

transient phenomena.—Actions which occur during the interval between two permanent or continually recurring conditions, such as slight changes taking place between peaks of an alternating current, changes during the charging of a condenser, etc.

transistor.—A device consisting essentially of a crystal, usually germanium of controlled impurity, in which some atoms have an excess of valence electrons while others lack the number of electrons required for stability. Excess electrons constitute negative charges which move through the crystal upon application of suitable potential differences at terminals. Where an electron is lacking there is a positive charge called a hole. The hole disappears from one position when that atom picks up an electron, only to reappear at the atom which lost the electron. Thus the positive charges or holes move toward a terminal which is made negative, while electrons move simultaneously toward a terminal made positive by the

TRANSIT TIME

applied potential difference. Transistors are used for amplification, oscillation, and for many other functions which might be served by electronic tubes. Transistor elements are called base, emitter, collector, and other names according to function. Basic types are junction transistors and point-contact transistors.

transit time.—Time required for electrons to travel from cathode to anode or plate in an electron tube, or for electrons and holes to move between terminals of a transistor crystal.

transition loss.—The *transmission loss* which would be eliminated by the use of a loss-less *passive transducer* at a junction point in a line. *Transformer loss*.

translucent.—Partially transparent; allowing passage of light but not permitting clear vision.

transmission efficiency.—The ratio of the power delivered from a transmission line or other apparatus to the power put into the line or apparatus.

transmission level.—The signal power or the *radio field intensity* at some point in a communication system, either in relation to some reference value or in absolute units of measurement.

Transmission Line (Equivalent)

transmission line.—Two insulated conductors having uniformly distributed inductance and capacitance, and characteristic impedance which depends chiefly on conductor spacing and insulation and is independent of either the length of line or the frequency of applied voltage and current carried. One of the many uses for transmission line is to conduct signals from antennas to receivers while providing impedance matching.

transmission loss.—The loss of energy suffered by a wave while passing through any circuit, any instrument, or other carrying medium. The transmission loss in any given part is equal to the ratio of the power which would be transmitted without the part in circuit to the power transmitted with the part acually in circuit.

transmission system.—A system of conductors carrying current which is of different form from that finally used by the power consuming devices; or a system of conductors carrying energy which is translated into a different form at the receiving end or load end of the circuits.

transmission unit.—Any unit which gives the logarithmic ratio of currents, voltages or powers in a transmission system. A *bel* (or *decibel*) and a *neper* (or *decineper*) are examples of transmission units. One transmission unit, in the usual meaning of the term, is the same as one decibel.

transmitter.—1—A *radio transmitter*. 2—A *microphone*.

transmitting tube.—A vacuum tube especially designed for use in radio transmitting circuits, generally a tube of large power rating.

transparent.—Permitting passage of light rays in such manner as to allow clear and distinct vision.

transposition.—Changing the relative positions of line wires at intervals to reduce or prevent magnetic, inductive and electrostatic unbalancing and disturbances to nearby electric devices.

transrectification.—*Rectification* occurring in the circuit of one tube electrode when voltage is applied to a different electrode.

transverse piezo-electric effect.—Electrical polarization and mechanical extension at right angles in a *piezo-electric crystal*.

transverse vibration.—1—Motion of the disturbed particles forming a wave in a direction at right angles to the direction of wave propagation. A characteristic of radio waves, light waves and the waves of radiant heat. Compare *longitudinal vibration*. 2.—A name sometimes applied to *flexural vibration*.

trap circuit.—A *wave trap*.

tree circuit.—A main or feeder circuit from which other circuits branch off like the limbs of a tree.

TRF.—Abbreviation for *tuned radio frequency*.

triad.—A trio.

tricolor picture tube.—A three-gun color television picture tube.

triggering.—Application to an oscillator, usually a sweep oscillator, of a sync pulse or other voltage which causes the oscillator to begin an operating cycle.

trimmer capacitor.—An aligning condenser or capacitor.

trio.—A triangular group of the three primary-color phosphor dots on the viewing screen or phosphor plate of a three-gun color television picture tube.

triode.—A three-element tube.

triode transistor.—A transistor having three elements, a base, an emitter, and a collector.

triple circuit jack.—A *jack* which, in addition to allowing connection to the plate circuit of an amplifying tube, also controls the filament current for tubes in following stages. *See illustration*.

triplex cable.—Three insulated conductors twisted together.

true focus.—The *real focus*.

true inductance.—A value of inductance which has been corrected from that of apparent inductance to take into consideration the *distributed capacity* of the circuit in its effect on frequency of resonance. Compare *apparent inductance*.

true power.—The effective power in an alternating current circuit; the power measured in *watts* as distinguished from the apparent power in *volt-amperes*. The number of volt-amperes multiplied by the *power factor*.

T-section.—A part of an electric circuit having two similar elements in series with one side of the circuit and another element in parallel with the circuit and connected between the two series elements. *See illustration*.

T.U.—Abbreviation for *transmission unit*.

tube.—1—A *vacuum tube*, a gaseous tube, a *photocell*, etc. 2—Tube of force—A *line of force*.

tube adapter.—A device which allows a tube of one type to be used in a socket intended for a different type.

tube base.—An insulating support carrying the glass bulb of a tube and having extended metal prongs which connect to the tube's internal elements.

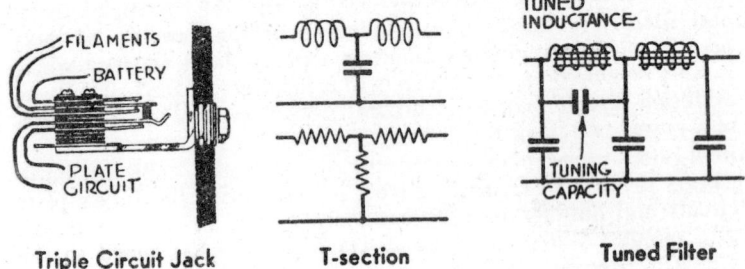

Triple Circuit Jack   T-section   Tuned Filter

tube capacities.—The electrostatic capacities between the elements of a vacuum tube. See *grid-plate capacitance, grid-filament capacitance, plate-filament capacitance;* also *grid capacitance, plate capacitance* and *filament capacitance*.

tube checker.—An instrument which indicates some of the simple characteristics of a vacuum tube; such as filament emission and the control of plate current by change of grid voltage.

tube impedance.—The *plate resistance* or plate impedance of a vacuum tube.

tube noise.—Noise due to amplification of slight variations of tube plate current such as produce noise-type voltages. Among principal causes are random irregularities at which electrons enter the plate, any other minute irregularities in electron flow, and leakages from control grid to other elements.

tube socket.—A device for supporting a tube while allowing circuit connections to be made with the prongs on the tube base.

tube tester.—A device for measurement of vacuum tube characteristics, such as *mutual conductance, plate resistance* and *amplification coefficient*.

tubular transmission line.—Transmission line whose conductors are embedded in opposite sides of dielectric insulation molded in the form of a hollow cylinder or tube.

tuned.—Descriptive of an *oscillatory circuit* in which the capacity, the inductance or both have been adjusted in value to cause *resonance* at some certain frequency.

tuned antenna.—An antenna circuit which is made *resonant* at the frequency to be transmitted or received by adjustment of capacity or inductance.

tuned audio frequency amplifier.—An audio frequency coupling system in which certain frequencies or certain frequency ranges are emphasized by using in these circuits values of inductance and capacity which produce *series resonance*.

tuned circuit.—An *oscillatory circuit*. A circuit containing capacity and inductance of such values as to produce *resonance*. Usually either the capacity or the inductance is variable in order that the resonance frequency may be changed at will.

tuned filter.—A *band exclusion filter* consisting of one or more *parallel resonance* circuits tuned to a particular frequency which is most troublesome, and which is thus most attenuated. The band exclusion circuit may be incorporated as part of a *low pass filter* in a power unit. *See illustration, page preceding.*

tuned-grid, tuned-plate oscillator.—A vacuum tube oscillator having a *parallel resonance* circuit in series with the tube's plate circuit and another parallel resonance circuit in series with the grid circuit. With both parallel resonance circuits tuned to the operating frequency, oscillation is maintained by capacitive feedback through the tube's internal capacity and also by inductive coupling in some instances.

tuned impedance amplification.—A radio frequency amplifying system in which the plate circuit of the amplifying tube contains a *parallel resonance* circuit tuned to the frequency to be amplified, thus producing maximum plate impedance and maximum amplification at this frequency.

tuned radio frequency amplification.—Voltage amplification at radio frequencies by means of vacuum tubes, the coupling being provided by means of air-core transformers having either tuned or aperiodic primaries and having secondaries tuned to *resonance* by means of variable condensers for the frequency of a signal to be received. Abbreviated *TRF*.

tuned transformer.—A *radio frequency transformer* having its secondary, its primary, or both secondary and primary windings tuned to *resonance* with the frequency applied to the primary. A similarly tuned *audio frequency transformer* or *intermediate frequency transformer*.

TUNER

tuner.—The portion of a receiver in which are circuits adjusted for resonance at carrier frequencies, most often with an associated radio-frequency amplifier or amplifiers. The channel selector circuits. In superheterodyne sound radio and television receivers the tuner ordinarily includes one or more radio-frequency amplifiers, an oscillator, and the converter or mixer, together with antenna coupling circuits, amplifier to mixer couplings, and oscillator circuits, all adjusted for resonance at frequencies suitable for reception and for formation of the intermediate frequency.

Tungar rectifier.—A trade name for an *argon rectifier*.

tungsten.—A rare metal used in the manufacture of vacuum tube filaments and heaters, also for switch contacts.

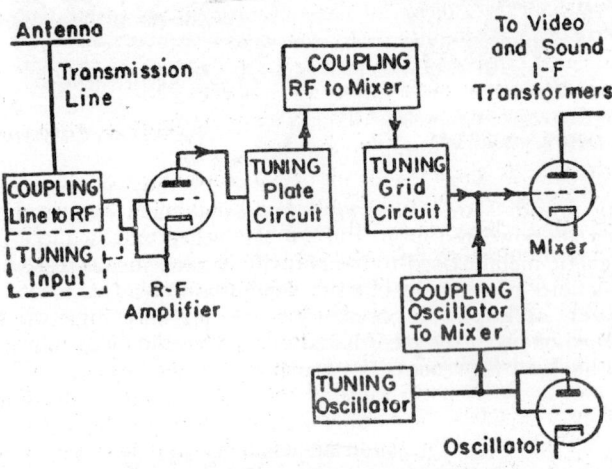

Tuner Elements

tuning.—Variation of the capacity or inductance in an alternating current circuit to cause *resonance* at a certain frequency called the tuned frequency, and thus to secure maximum power in reception or transmission of signals at this frequency.

tuning coil.—An inductance coil used in a circuit which may be tuned to *resonance*.

tuning condenser.—A *variable condenser*, generally of the air dielectric type, which is used as the adjustable part of a circuit to be tuned to *resonance*. Also, several such condensers arranged for simultaneous operation from one control. *See illustration.*

tuning dial.—A device which indicates the setting of a tuning inductance or capacity either in kilocycles, wavelengths or on an arbitrarily numbered scale.

TUNING FORK OSCILLATOR

tuning fork oscillator.—An *audio frequency oscillator* in which the frequency is determined by vibrations of a metallic tuning fork. The circuit for supply current passes through a *microphone button* attached to one tine of the fork. *See illustration.*

tuning inductance.—A *tuning coil.*

tuning note.—A characteristic audio signal or tone sent out from a transmitter to allow *tuning* of receivers before a program.

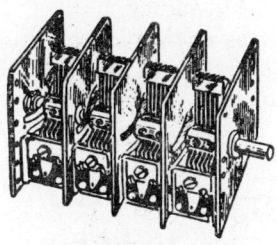
Tuning Condenser

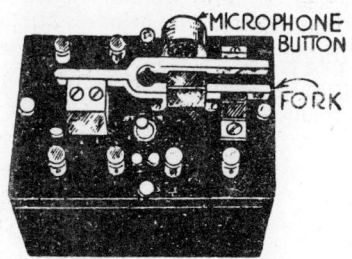
Tuning Fork Oscillator

tuning wand.—An insulating support having at one end a small piece of powdered iron and at the other end a piece of non-magnetic metal. Used to temporarily increase or decrease apparent inductance of resonant circuits during service adjustments.

turnover pickup.—A phonograph pickup carrying two needles. Turned one way for records requiring a needle of certain form and another way for records requiring a different form.

Turntable

turns ratio.—The ratio of the number of turns in a high-voltage winding to the number of turns in the low-voltage winding of a *transformer.*

turntable.—That portion of a phonograph apparatus which supports and rotates the disc record as it passes underneath the pickup needle. *See illustration.*

TURRET TUNER

turret tuner.—A tuner, most often in a television receiver, having a separate set of resonating circuit elements for each channel, with each set mounted on an insulating strip or strips carried by a cylindrical drum rotated from the channel selector control. At each drum position the terminals of circuit elements for one channel make contact with terminals leading to tubes and other parts mounted on the stationary frame of the tuner.

tweeter.—In an assembly of two or more speakers, one which is of such size and design as to more effectively reproduce the higher audio frequencies.

twin-diode or -triode.—A tube having a single envelope within which are two independent sets of elements.

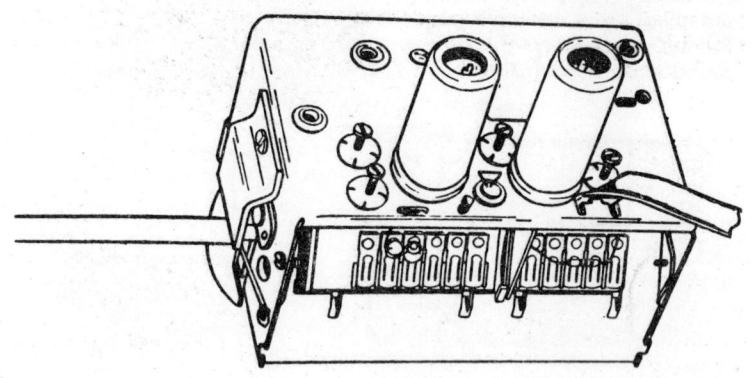

Turret Tuner

twin speakers.—Two loud speakers connected to a single amplifier; one speaker being adapted to reproduction of low audio frequencies and the other to reproduction of higher frequencies.

twisted pair.—Two separate insulated conductors twisted together but having no common covering.

two-band antenna.—A television antenna consisting of one set of elements for reception of low-band signals and a separate set for high-band signals.

two-element tube.—A diode tube.

two-phase.—Descriptive of an alternating current system in which there are two separate e.m.fs. and two currents at one time, the angle between phases being 90 degrees.

two-pole switch.—A *double-pole switch*.

two-way switch.—A *double-throw switch*.

two-wire circuit.—A metallic circuit using paralleled conductors.

U

UHF or uhf.—Abbreviation for ultra-high frequency.

ultor.—The electrode of a television picture tube operated at highest voltage, otherwise called the anode or second anode.

ultraudion circuit.—A *regenerative detector* system in which a tuned *parallel resonance* circuit is connected between the grid and plate of the vacuum tube, a blocking condenser being placed between the grid and the high voltage circuit. Regeneration is controlled by a variable condenser between the tube's plate and cathode. Increasing the reactance of this condenser increases the feedback and the intensity of regeneration. *See illustration.*

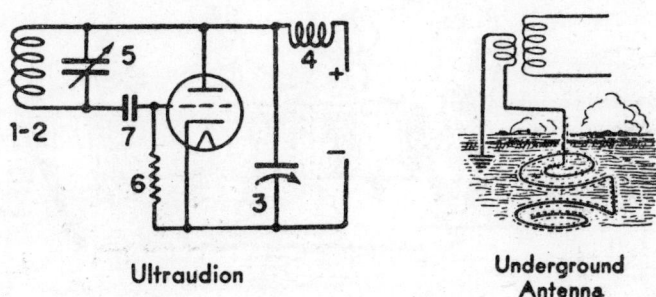

Ultraudion   Underground Antenna

ultra-high frequency.—A frequency in the range from 300 to 3,000 megacycles per second. For television broadcasting the uhf range is presently from 470 to 890 megacycles, within which are channels 14 to 83 inclusive.

ultra-violet rays.—*Radiant energy* at higher frequencies or shorter wavelengths than those which compose *visible light*. Rays lying between those which are visible and the lower frequency limits of X-rays.

unbalanced line.—A coaxial line with its inner conductor connected to one side of a dipole antenna and the outer conductor or braid connected to the other side of the antenna and to ground.

umbrella antenna.—An antenna having a number of aerial conductors extending outward and downward toward the earth from a central support.

undamped oscillations or waves.—Oscillations which continue with undiminishing amplitude because of energy supplied to overcome the circuit damping. *Continuous waves.*

underground antenna.—A receiving type of *capacity antenna* having one long insulated conductor buried in the earth or immersed in water, either in an extended or a coiled form. See *illustration*.

undistorted power.—*Maximum undistorted power output* of a vacuum tube.

undulatory theory.—The theory that *radiation* is propagated through space by wave-like disturbances.

unidirectional current.—A current (or voltage) always acting in one direction, not reversing, although it may vary in amplitude.

unidirectional direction finder.—A *direction finder* with which is combined a *sense finder*.

unifilar winding.—A winding using but a single conductor.

unilateral antenna.—An antenna which radiates or receives more effectively in one direction than in any other direction.

unilateral conductivity.—Having less resistance to flow of current in one direction than to flow in the opposite direction; the chief property of a *rectifier*.

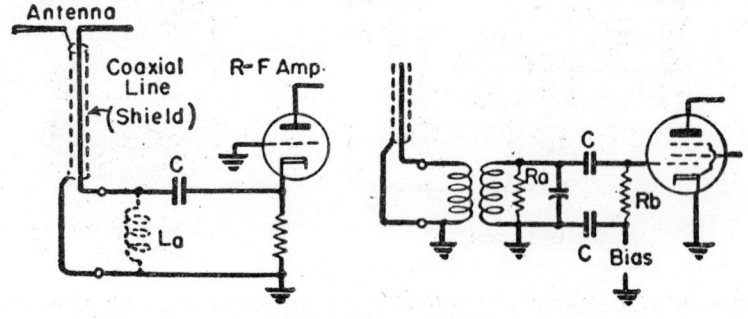

Unbalanced Line

unit charge.—The quantity of electricity which exerts a force of one dyne upon an equal quantity one centimeter distant.

unit flux density.—A *magnetic flux density* of one *gauss*.

unit force.—A *force* of one *dyne*.

unit magnetic field.—A *magnetic flux density* of one *gauss*.

unit magnetic flux.—A flux of one *maxwell* or one *line of force*.

unit magnetomotive force.—A *magnetomotive force* of one *gilbert*.

unit permeability.—The *permeability* of non-magnetic material, which is equal to 1.

unit pole.—A magnetic pole of such strength that when placed in a vacuum one *centimeter* from an equal pole it repels the other pole with a force of one *dyne*.

unit pressure.—A pressure of one *dyne* per square centimeter.

unit reluctance.—A *reluctance* of one *oersted*.
unit surface.—A surface of one square centimeter.
unit velocity.—A velocity of one centimeter per second.
unit volume.—A volume of one cubic centimeter.
unity power factor.—A *power factor* of 1.0, which exists when the current and voltage are in phase. The condition in a circuit containing resistance but no *reactance*, or the condition at *resonance* in a reactive circuit.
universal motor.—A motor which operates either on direct or alternating current. The construction is like that of a series wound direct current motor.
untuned radio frequency transformer.—A coupling transformer operating over a wide range of radio frequencies but not *tuned* to resonance at any particular frequency.
u. p. o.—Abbreviation for *undistorted power output*.
upper side band.—The *side band* frequency which is higher than the *carrier frequency*.
uranium.—A metallic element which is a *radioactive substance*.

V

V.—Symbol for *potential difference*.
v.—Symbol for *velocity*, also for *volts*.
vacuum circuit breaker.—A *circuit breaker* the contacts of which are located in a vacuum for the purpose of lessening the arc.
vacuum phototube.—A phototube in whose envelope or bulb the vacuum is sufficiently complete that there is no intentional ionization during operation. Compare gas phototube.
vacuum thermocouple.—A *thermocouple* and heating element enclosed within an evacuated chamber for the purpose of increasing the sensitivity through a decrease in the rate of heat radiation from the element. *See illustration.*
vacuum tube.—A glass bulb from which gases have been almost completely exhausted and within which are various electrodes required for operation of the tube as an amplifier, a rectifier, a modulator, an oscillator, a detector, etc. Action of a vacuum tube is due entirely to *thermionic* action or electron flow through the evacuated space between electrodes.
vacuum tube amplifier.—A radio frequency or audio frequency amplifier utilizing the amplification of *vacuum tubes* to increase the signal voltage or power.
vacuum tube detector.—A detector employing a vacuum tube for *grid current detection*, for *plate current detection*, or for detection as a diode.
vacuum tube generator.—A *vacuum tube oscillator*.
vacuum tube oscillator.—A device employing a vacuum tube using direct current power for the production of alternating currents at desired frequencies.
vacuum tube rectifier.—A *thermionic rectifier*.
vacuum tube relay.—An oscillating tube with a control circuit connected to its control grid; increase of grid voltage resulting in an increase of plate current sufficient to operate an electromagnetic *relay* in the plate circuit.
vacuum tube voltmeter.—An instrument with which voltages to be measured are applied, usually through a voltage divider or attenuator, to the grid-cathode circuit of one or more tubes whose plate or cathode currents actuate a moving coil meter graduated to read applied voltages. Since a negatively biased grid carries no appreciable current, the grid cathode input circuits may be of high resistance or impedance for even the lowest measured voltages. This allows taking little power from measured circuits and provides high sensitivity of the meter.

valence electron.—An electron loosely bound in the outer shell of an atom and giving that atom one more or one less electron than required for stability. A valence electron easily leaves an atom where that electron is in excess over a stable number, and passes to any other atom where it may complete a stable shell. Valence refers to the number of electrons which must be gained or lost to make an atom stable. Atoms may be bonded together by having in common one or more electrons which would be an excess for one of the atoms and a deficiency for the other atom.

valve effect.—Any effect which offers greater opposition to flow of electricity in one direction than in the opposite direction.

valve metal.—A metal which, when immersed in an electrolyte, allows electric current to flow in only one direction between the metal and the liquid. A metal used as the active electrode in an *electrolytic rectifier*.

vane type instrument.—1—A current measuring instrument

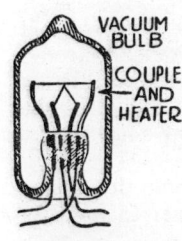

Vacuum Thermocouple

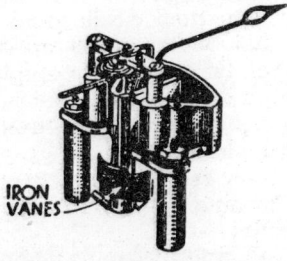

Vane Type Instrument

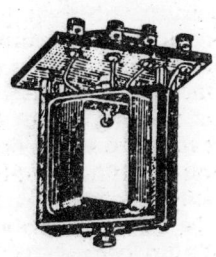

Variometer

employing the repulsion force between fixed and movable magnetized iron vanes to move the pointer. 2—An instrument employing the reaction between the field of a coil carrying the measured current and a magnetized piece of soft iron. *See illustration.*

V-antenna.—An *antenna* having two aerial conductors extending in different directions from a central support.

variable area system.—A method of sound recording on film in which part of the total width of *sound track* is opaque and the remainder transparent, an irregular line of separation between these parts representing variations of sound frequency and intensity.

variable condenser.—Any condenser in which the effective *capacity* may be continuously changed while operating in a circuit.

variable density system.—A method of sound recording on film in which alternate light and dark lines across the *sound track* represent the sounds. The number of lines per unit of film length is proportional to frequency and the contrast between light and dark is proportional to intensity of the sound.

VARIABLE-MU TUBE

variable-mu tube.—A pentode having remote cutoff.

variable pulse width afc.—An automatic frequency control for television sweep oscillators. Oscillator grid voltage and operating frequency are altered by the charges imparted to capacitors. The charges are formed by pulses of current from the cathode side of a control tube made conductive only during sync pulse periods and to an extent dependent on the time relation between sync pulses and the sharp slope of a sawtooth wave derived from the oscillator output. Variation of oscillator frequency shifts the phase of the sawtooth with respect to sync pulses, to change the time duration or width of charging pulses. Changes of capacitor charge and voltage act on the oscillator grid to correct its frequency.

variable reluctance cartridge.—A phonograph pickup in which needle movement varies the reluctance and magnetic flux in a circuit, thus inducing changes of e.m.f. and current at sound frequencies.

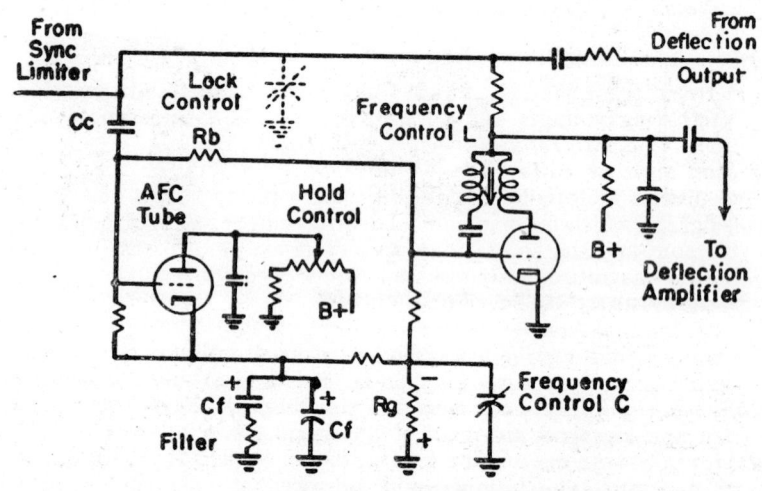

Variable Pulse Width Afc

variocoupler.—A radio frequency *inductive coupling* device in which the *coupling coefficient* may be changed by relative movement between two windings or by using more or less of the turns in one of the windings.

variometer.—A continuously variable *inductance* consisting of two windings, one of which may be rotated inside the other so that the two fields either aid or oppose each other to vary the total inductance. *See illustration, page preceding.*

varistor.—A resistor having a large negative temperature coefficient.

VARNISHED CLOTH

varnished cloth.—Linen or cotton coated with oils or resins.
vector.—A symbol for a changing quantity which, at a given instant, has definite direction and magnitude. A straight line representing by its length the magnitude of an alternating quantity and by its angular position in relation to other lines the *phase* relation

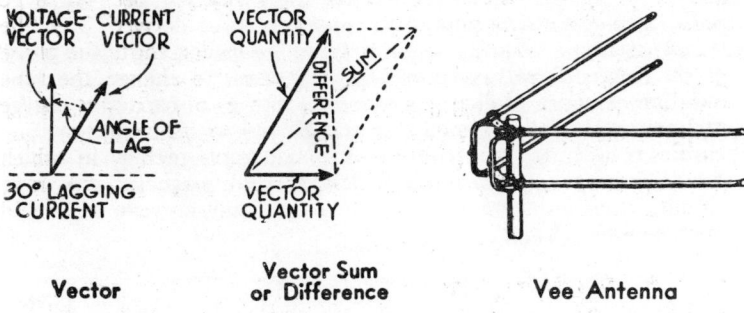

Vector

Vector Sum or Difference

Vee Antenna

between this quantity and others. Thus it is possible to represent alternating voltages and currents and to show their phase relations. Compare *rotating vector*. *See illustration.*
vector sum or difference.—The result of adding or subtracting quantities vectorially. Two *vectors* form the two sides of a parallelogram, which is completed. Then the longer diagonal represents the sum and the shorter one the difference of the two quantities both in magnitude and in phase relation. *See illustration.*
vee-beam antenna.—A dipole antenna whose conductors spread from the signal takeoff gap in the form of a letter V on a horizontal plane, with the wide opening toward the direction from which desired signals approach.
velocity.—The rate of movement, the distance traveled in a given direction during a given time. The symbol is v.
velocity constant or factor.—The rate at which waves of voltage or current travel in a transmission line or other inductive-capacitive path, expressed as a fraction or a percentage of the rate in free space.
vernier condenser.—A *tuning condenser* in which the capacity may be changed by a very small amount through an auxiliary control acting upon a small portion of the total capacity.
vernier dial.—A device by means of which the shaft of any variable unit may be turned through a very small angle of rotation or may be turned very slowly with the control knob turned comparatively fast.
vertical blanking.—The time interval between successive fields of television pictures during which composite signal voltage remains at the black level except for blacker than black vertical and

equalizing sync pulses. The electron beam is presumed to be cut off. Vertical retrace occurs during this period.

vertical polarization.—Descriptive of a carrier wave whose electrostatic lines are approximately vertical or perpendicular to the surface of the earth, while magnetic lines are approximately horizontal.

vertical recording.—A method of phonograph recording with which the groove on the disc is of constant width but varies in depth to cause up and down motion of the needle.

vertical retrace.—During vertical blanking periods between television picture fields, continuation of sawtooth voltages and currents which cause vertical deflection for pictures. During blanking the composite television signal is at the black level or blacker than black for sync pulses, and there should be no electron beam. Were the beam to exist it would "retrace" a zig-zag upward path.

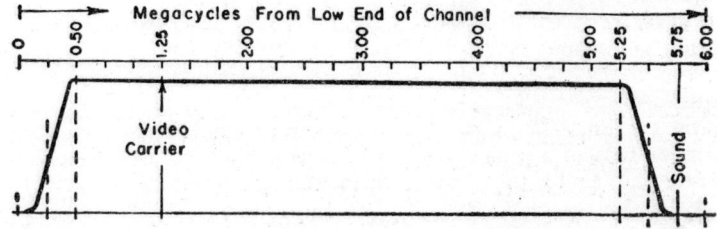

Vestigial Sideband Transmission

vertical retrace lines.—Bright, widely spaced sloping lines which appear on television pictures when the electron beam continues during vertical blanking periods.

very-high frequencies.—The range of frequencies from 30 to 300 megacycles per second. Television broadcast carrier frequencies for channels 2 through 13 are in this range, between 54 and 216 megacycles.

vestigial sideband transmission.—The standard method of television carrier transmission. Signal modulation includes video carrier frequency and all upper sidebands extending to four or more megacycles above the carrier, but lower side bands extending only to 0.75 megacycle below the carrier. The upper and lower side bands are transmitted within the ranges to 0.75 megacycle either way from the carrier, but only the upper sideband is transmitted for all greater frequencies. Compare receiver attenuation.

VHF or vhf.—Abbreviation for very-high frequency.

video.—A word which refers to television pictures and to any circuits or components which assist in transmission, reception or reproduction of pictures.

VIDEO AMPLIFIER

video amplifier.—Amplifier tubes and associated circuits primarily intended to strengthen television picture signals between the video detector and picture tube, but which ordinarily carry sync and blanking signals as well.

video carrier.—The television signal whose modulation sidebands include all picture, sync and blanking signals. The carrier itself is at a frequency 1.25 megacycles higher than the low limit of the channel, and 4.5 megacycles lower than the sound carrier in the same channel.

video detector.—A television receiver demodulator, most often a diode tube or a crystal diode, whose function is to recover from amplitude-modulated intermediate frequency voltages the signals for pictures, sync pulses and blanking.

video frequencies.—A range of television signal frequencies extending from as low as 30 cycles per second to a high limit of something less than 4.5 megacycles per second, within which are frequencies for all picture signals and sync pulses.

video tape recording.—Recording and reproduction of television picture and sound signals on tape similar to that commonly used for sound recording. The wide range of video frequencies may require several tracks on one tape.

viewing screen.—The inner surface of a television picture tube face plate, or that of a separate internal plate, on which are phosphors made luminous by the electron beam during formation of pictures or raster.

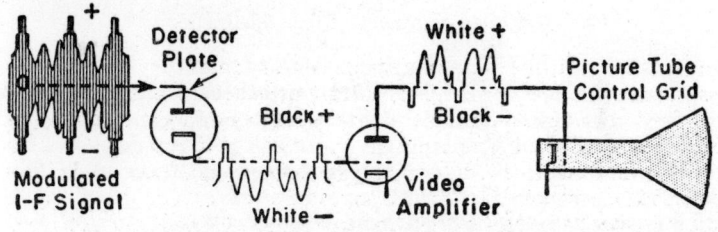

Video Detector and Amplifier

violet rays.—A term usually meaning *ultra-violet rays*.

virtual focus.—The point at which imaginary extensions of diverging rays from a *lens* would meet were they extended back through the lens. *See illustration*.

virtual image.—An image formed by *reflection* or *refraction*. Such an image as appears to the eye but which cannot be focused on a screen. *See illustration*.

virtual value.—An effective value, or *root-mean-square value*.

visible light.—Light rays of such wavelength or such color as will

VISIBLE SPECTRUM

affect normal vision; distinguished from *ultra-violet rays* and from *infra-red rays* which are not visible. Wavelengths between 3,900 and 7,600 *Angstrom units* and including the colors violet, indigo, blue, green, yellow, orange and red.

visible spectrum.—Light frequencies or wavelengths within the limits of human vision. See *spectrum*.

vision.—Sight; the sense which is affected by light, color and form.

visual.—Pertaining to the sense of sight, or to vision.

visual angle.—The angle formed by two straight lines drawn from the outer limits of an object or image, and meeting in the eye. See *illustration*.

visual field.—The portion of an image which may be clearly seen without moving the eye.

visual persistence.—*Persistence of vision*.

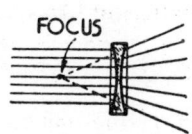 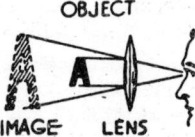 

Virtual Focus   **Virtual Image**   **Visual Angle**

vitreous.—Glasslike. Having the nature of glass.

voice coil.—A few turns of conductor attached to the small central extension of a speaker cone and thus supported in the magnet gap. Audio-frequency currents in the voice coil produce magnetic fields which react with the field of the stationary magnet to cause vibration of the coil and attached cone.

voice frequency.—*Audio frequency*.

voice modulation.—*Speech modulation*.

volt.—The practical unit of *electromotive force* or electrical potential. The steady electromotive force which causes a current of one *ampere* to flow through a resistance of one *ohm*. Abbreviated *v*. The *international volt*.

Volta effect.—Development of an *electromotive force* at the point of contact when different materials are brought together.

voltage.—A difference of potential between two points in a circuit, as measured in volts.

voltage amplification.—The ratio of the signal voltage appearing between the output terminals of an amplifier to the signal voltage applied to the input terminals. With a vacuum tube, the *amplification coefficient*.

VOLTAGE DIVIDER

voltage divider.—1—A resistance unit provided with one or more sliding contacts and having impressed across its ends a *potential difference*. Various voltage values may be had between a slider and either end of the resistance, or between two sliders. *See illustration.* 2.—That part of a *power unit* which consists of resistors through which flows all or part of the rectified current and from various points on which are taken potentials suitable for the plate circuits, screen circuits and control grid biases of the connected apparatus.

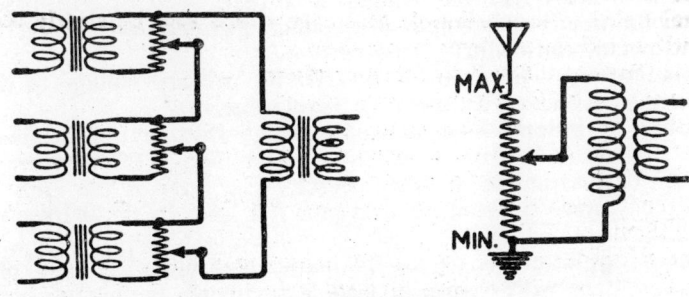

Voltage Dividers

voltage doubler.—A voltage multiplier whose d.c. output voltage is approximately twice the r-m-s value of alternating input voltage.

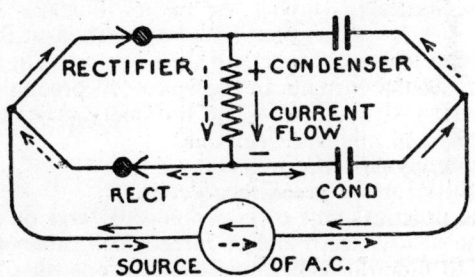

Voltage Doubler

voltage drop.—*Potential difference.* The difference between the voltages at two points in a circuit.
voltage factor.—The *amplification coefficient*.
voltage feed.—A connection from the closed circuits of a *radio transmitter* to its antenna at a point where the antenna has a high voltage but carries little current.

voltage feedback.—A method of degeneration with which feedback voltages whose phase is opposite to that of voltages at the input are obtained from some circuit following the output of the degenerative amplifier.

voltage multiplier.—A circuit containing two or more rectifiers and several capacitors so connected as to charge in parallel or separately from rectifier d.c. output currents, but to discharge in series to add together the capacitor voltages.

voltage operated device.—An electrical part the operation of which depends on changes in applied voltage or potential differences rather than on changes in current. A negatively biased vacuum tube is an example of a voltage operated device since the grid circuit carries little or no current.

voltage ratio.—The ratio of the effective primary voltage to the effective secondary voltage in a transformer.

voltage regulation.—The change in voltage which is brought about by change of load on a generator, a rectifier, a transformer or other unit acting as a power source. The ratio of the drop in output voltage between no load and full load, to the voltage at full load or rated load.

voltage regulator.—A device for maintaining approximately constant voltage with change of load in a circuit. Various methods are employed, as a transformer with variable mutual inductance. See also, *voltage regulator tube*.

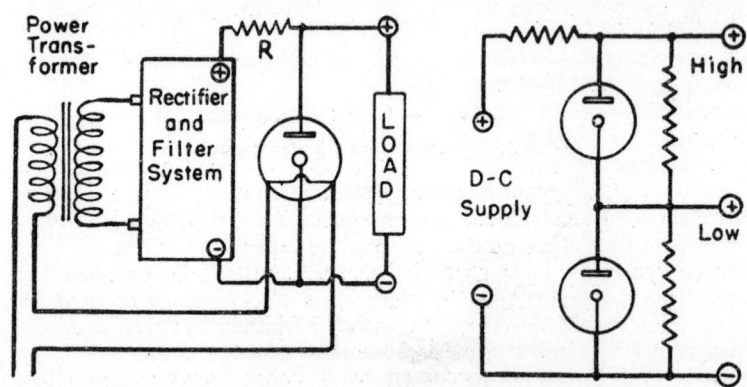

Voltage Regulator Tube

voltage regulator tube.—A *two-element gaseous tube* in which a *glow discharge* occurs and in which the voltage drop between elements varies with change in external load to maintain an approximately constant voltage between points to which the tube is connected.

voltage relay.—A relay which operates upon a change of applied voltage.

voltage resonance.—*Series resonance.*

voltage transformer.—A transformer having its primary connected between the two sides of a circuit, primary current then depending on the *potential difference* across the circuit.

voltaic cell.—*A galvanic cell.*

voltaic couple.—Two substances (usually metals) between which there is developed an *electromotive force* when they are placed in an *electrolyte*.

voltaic electricity.—The *electric current*.

voltaic pile.—Alternate discs of two different metals between which are layers of absorbent material carrying *electrolyte*, the combination acting as a number of *primary cells* in series.

voltammeter.—1—An instrument combining two movements, for simultaneous measurement of both voltage and current in a circuit, or containing a single movement with switching arrangements for measuring either voltage or current. 2—A direct current *wattmeter*.

volt-ampere.—A unit of measurement for apparent power in an alternating current circuit as distinguished from real power which is measured in *watts*. The number of volt-amperes is equal to the product of the r-m-s applied volts and the r-m-s current in amperes in a circuit, without reference to the *phase* relation of voltage and current or to the *power factor* of the circuit. In a direct current circuit the number of volt-amperes is the same as the number of watts.

volt-ampere ratio.—The ratio of the *volt-ampere* output to the volt-ampere input of a transformer.

Volta's law.—A law which states that the voltage across the ends of a *series circuit* containing several sources is equal to the sum of all separate e.m.fs. in one direction minus the sum of all separate e.m.fs. acting in the opposite direction.

voltmeter.—An instrument for measuring and indicating *potential difference* in volts.

volts to ground.—In a grounded circuit, the voltage between the specified conductor and the connection of its circuit to ground. In an ungrounded circuit, the greatest voltage between a specified conductor and any other conductor of the circuit.

volume.—1—The space occupied by a body; measured in cubic feet, cubic inches, etc. 2—The intensity of sound produced by a loud speaker.

volume compression.—Manual or automatic reduction of the variation or range of audio power between its minimum and maximum volume levels at the input of a recording or transmitting system, with respect to the volume range of original sound at the input.

volume control.—Manual or automatic means for regulating the volume of sound produced by an audio frequency amplifier in its connected *loud speaker*.

volume distortion.—*Amplitude distortion.*

volume expansion.—The inverse of volume compression. An increase of variation or range between minimum and maximum volume levels at the output of an audio amplifying system in comparison with the volume range represented by signal input to the system.

volume indicator.—A device which visually indicates the amount of power, current or voltage being carried by a transmission line. One form consists of a simple *vacuum tube voltmeter*, other types employ direct reading alternating current meters. *See illustration.*

volume leakage.—The *current* which passes through the body of an insulating material.

volume resistance.—*Resistance* to flow of current through the body of a material.

volume resistivity.—The *resistance* in ohms between opposite faces of a centimeter cube of a substance. Sometimes the resistance in ohms of one mil-foot of the substance.

volume unit.—A unit for measurement of audio-frequency power in complex waves, in much the same way that the decibel is used for sine waves.

VT.—Abbreviation for vacuum tube.

VTVM.—Abbreviation for vacuum tube voltmeter.

vulcanite.—*Hard rubber.*

vulcanized fibre.—Fibre which has been formed with application of heat.

W

W.—Symbol for *work;* for energy in *joules*.

w.—Abbreviation for *watt*.

Wagner ground.—Auxiliary variable arms connected to opposite corners of a *Wheatstone bridge* and to ground; allowing elimination of the effects of stray capacities and of leakages. *See illustration.*

wander plug.—A metallic plug attached to a flexible wire, through which connection may be made to any one of several points in a circuit.

watchcase receiver.—A small *telephone receiver*. A *headphone*.

watt.—The practical unit of electric *power*. A power of one *joule* per second. The power produced in a direct current circuit by a current of one ampere through a potential drop of one volt, the power in watts being equal to the product of the number of amperes flowing and the voltage drop across the circuit. In an alternating current circuit the (true) power in watts is equal to the number of *volt-amperes* multiplied by the *power factor* of the circuit. Abbreviated *w*.

wattful current or component.—The *active current or component*.

watt-hour.—The amount of power furnished by one *watt* when used for one hour.

watt-hour meter.—An integrating wattmeter; a device for recording the total power in *watt-hours* used in a circuit.

wattless current or component.—The *reactive current or component*.

wattmeter.—An instrument for measuring and indicating in *watts* the rate at which *power* is used in a circuit. In one type the pointer is moved by the reaction between two windings, one in series and the other in shunt with the circuit in which power is being measured. *See illustration.*

watt-second.—The amount of power furnished by one *watt* during one second. One *joule* of energy.

wave.—A moving disturbance in an elastic body, the disturbance having a regularly recurring period in time. A single cycle of such a disturbance. The representation of such a wave in a graph. A *radio wave, light wave, sound wave,* etc.

wave antenna.—A directional receiving *antenna* in which the horizontal length of the aerial conductor is a simple multiple of the wavelength of the received carrier. The open end is at a voltage *node* and is connected through a resistance to ground.

wave band.—The series of wavelengths or frequencies within which the carrier of a transmitter is maintained. A *radio channel*.

wave filter.—A *filter*.

wave form.—The shape of a curve indicating the *instantaneous values* in the rise and fall of an alternating quantity during a period of time. The wave form of a signal is not dependent on the amplitude of voltage or current, nor on the scale to which it may be drawn on a graph; remaining the same in form after amplification as before provided there has been no *distortion*.

wave form distortion.—A form of distortion in which the shape of the signal wave is changed, usually by introducing *harmonic frequencies* not present in the original signal.

wave front.—The points in space at which the effects of one particular cycle of a moving disturbance have the same *phase*.

wave front angle.—The angle at which a radio wave deviates from

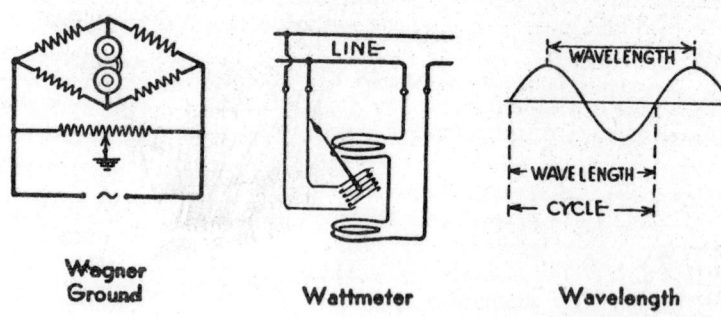

Wagner Ground   Wattmeter   Wavelength

the perpendicular because of *refraction* in the mediums through which it is traveling.

wavelength.—The distance between successive peaks of the same polarity in a *wave*. The distance traveled by one wave in space before a similar part of the following wave arrives at the same point. The distance spanned by one cycle of radiated energy. Equal to the wave *velocity* divided by the *frequency*. Wavelength usually is measured in *meters* or in *centimeters*, although for very short waves other units, such as the *Angstrom unit*, are employed. The symbol is the Greek letter lambda (λ). *See illustration.*

wave meter.—A device which allows measurement of the *wavelength* of an alternating current or of a radio wave. A *frequency meter* calibrated to measure wavelength.

wave motion.—A to and fro motion of an elastic medium as disturbances of any kind travel through it. The motion may be in the form of *longitudinal vibrations* or of *transverse vibrations*.

wave propagation.—The movement of waves through space.

wave reflection.—See *reflection*.

WAVE SELECTOR

wave selector.—A *wave trap*.

wave shape.—*Wave form*.

wave trap.—A series resonance circuit or a parallel resonance circuit employed to improve the selectivity of receiving apparatus against various kinds of interference, including undesired radio or television signals. A parallel resonance trap in series with a signal path attenuates the frequency for which the trap is tuned. A series resonance trap from a signal circuit to ground or to some circuit at ground potential for signal frequencies will divert to ground or the other circuit any frequency for which the trap is tuned.

wax master.—In the disc method of sound recording, the "wax" disc into which is cut the original record of sounds by the cutter. Used for producing the *metal master*.

weak coupling.—*Loose coupling*.

Wehnelt filament.—An *oxide coated filament*.

Wehnelt interrupter.—An *electrolytic interrupter*, consisting of one lead and one platinum electrode immersed in dilute sulphuric acid.

Wehnelt oscillator.—A generator of high frequency oscillations employing a *Wehnelt interrupter* connected in a circuit containing suitable values of inductance and capacity.

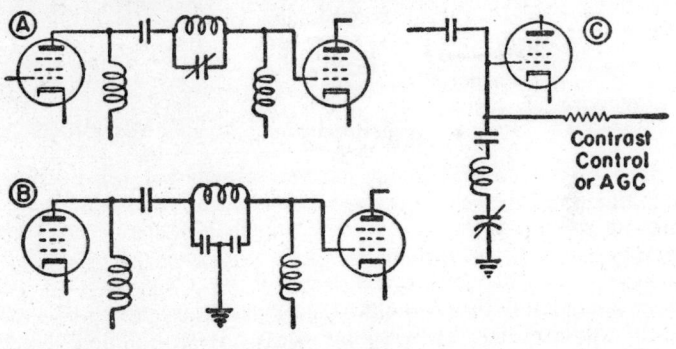

Wave Traps

Wein bridge.—A bridge circuit having a capacitor and resistor in series for one arm, a capacitor and resistor in parallel for another arm, and resistors for the remaining two arms. Used for measuring audio frequencies, also, with adjustable capacitors, as the frequency determining element for audio oscillators or signal generators.

Weston cell.—A *standard cell* having one electrode of mercury, another of cadmium amalgam and an electrolyte of mercurous sulphate and cadmium sulphate. Delivers an electromotive force of 1.01865 volt at a temperature of twenty degrees centigrade.

WET ELECTROLYTIC CAPACITOR

wet electrolytic capacitor.—An electrolytic capacitor whose electrolyte liquid is not held in absorbent material but is free within a can containing the electrodes. This type has somewhat more leakage than dry electrolytics, but tends to be self-healing after momentary breakdown of the dielectric.

Wheatstone bridge.—A device for measurement of electrical quantities. Four arms form a parallel circuit across a voltage source, two arms on each side, and a bridging member containing a current indicator joins the points between the two arms on either side. When the ratio between one pair of arms is made equal to the ratio between the other pair, the bridge is balanced and no current flows through the indicator. One ratio then is determined

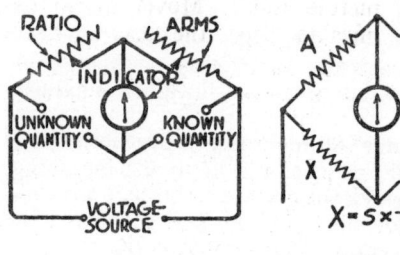

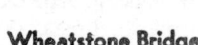

Wheatstone Bridge

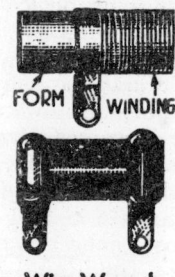

Wire Wound Resistor

from marked calibrations on its two arms; the other ratio is equal and consists of the unknown quantity and a known quantity, therefore the unknown may be determined in terms of the known quantity. *See illustration.*

white level.—In modulation consisting of the composite television signal, the modulation voltage corresponding to greatest brightness in pictures, usually at about 15 per cent of maximum amplitude. See negative transmission.

wide angle yoke.—A deflecting yoke for television picture tubes in which the deflection angle exceeds 62 to 65 degrees.

winding.—A portion of an electrical circuit formed into a coil.

wipe-out area.—A region in which *blanketing occurs.*

wire.—A slender rod or filament of drawn metal. One or more pieces or strands of metal not insulated from each other.

wire recording.—Recording of sound by imparting magnetization along an iron alloy wire in strengths and at intervals corresponding to audio frequency and amplitude, while the wire is drawn through a recording head. When the wire is run through a reproducer the magnetic variations induce audio-frequency currents.

wire skinner.—A tool having knife edges formed into openings the size of conductors, so that clamping the tool down on an insulated wire allows the insulation to be drawn off without cutting the conductor.

wire wound resistor.—A resistance element consisting of a metallic wire wound upon a heat resisting, insulating form and usually covered with enamel or other material to prevent oxidation. *See illustration.*

wired radio.—Radio transmission in which the modulated *carrier current* is guided by metallic conductors, these conductors usually being already in use for carrying power, or telephonic or telegraphic messages.

wireless.—A word having the same meaning as *"radio"*.

wobble plate.—The adjustable member of a permanent-magnet centering device for television picture tubes. Movement of the plate shifts the magnetic field lines to direct the electron beam as required.

wobbulator.—A sweep generator.

woofer.—In an assembly of two or more speakers, a unit designed for most effective reproduction of lower audio frequencies or of middle and low frequencies.

work.—The product of a *force* and the distance through which it moves a certain mass. Units of work are the *erg*, the *joule*, the foot-pound and the centimeter-gram. The symbol is W.

work function.—The least amount of energy, expressed in volts, which an *electron* must acquire in order to break away from the surface of a metal into the surrounding space. The potential difference in volts through which an electron must fall to gain velocity sufficient to allow its breaking away from the surface of a *cathode*. Surface tension.

working voltage.—With reference to capacitors, the maximum continuous direct voltage which may be applied without danger of dielectric breakdown.

X

X. or x.—Symbol for *reactance* in ohms.
X$_c$.—Symbol for *capacitive reactance* in ohms.
X$_l$.—Symbol for *inductive reactance* in ohms.
X-axis.—A line joining two diametrically opposite corners of a quartz *piezo-electric crystal*, and lying in a plane at right angles to the *Z-axis*. An electric axis. *See illustration.*

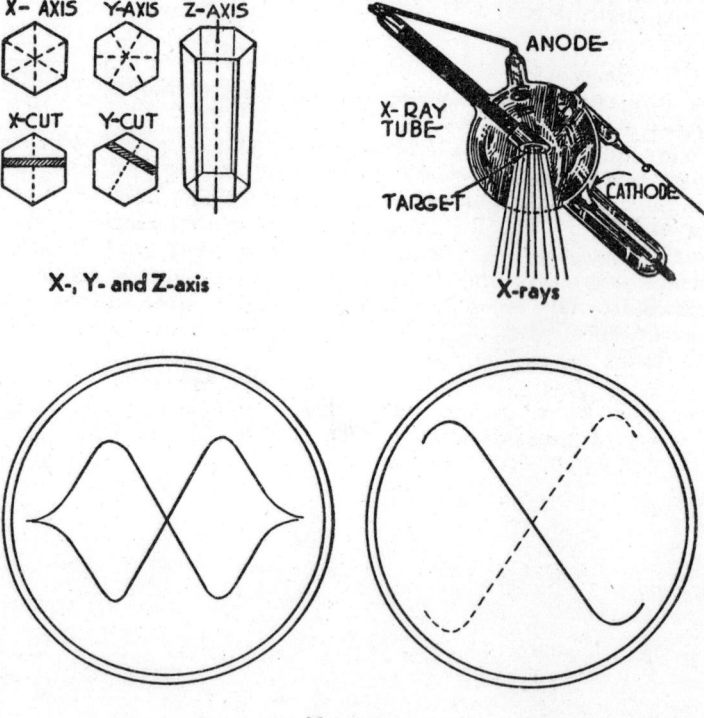

X-, Y- and Z-axis

X-curve

X-curve.—A curve or trace of audio-frequency output voltage from a frequency-modulation demodulator. The form is that of two lines crossing somewhat in the manner of a letter X. Such a curve appears when horizontal sweep rate in the oscilloscope is twice that at which frequency deviates from a center frequency in the sweep generator.

X-cut.—Descriptive of a *piezo-electric crystal* or quartz plate cut in such manner that an *X-axis* is perpendicular to its faces. Also called a Curie cut, a zero-angle cut or a face-perpendicular cut. *See illustration.*

X-rays.—Rays produced by striking of *cathode rays* on a solid. These rays are capable of penetrating opaque objects and will affect photographic plates or will produce *fluorescence*. Rays having frequencies lying approximately between the higher ultraviolet frequencies and the lower gamma rays. *See illustration.*

X's.—*Static* disturbances.

X-waves.—In a *piezo-electric resonator,* waves whose direction is parallel to an *X-axis*.

Y

Y. or y.—Symbol for *admittance* in mhos.
Yagi antenna.—A half-wave dipole antenna used with at least two and usually with four or more directors, and with a single reflector. Provides high gain, sharp directional properties, but response in a limited frequency range, often for only one channel.

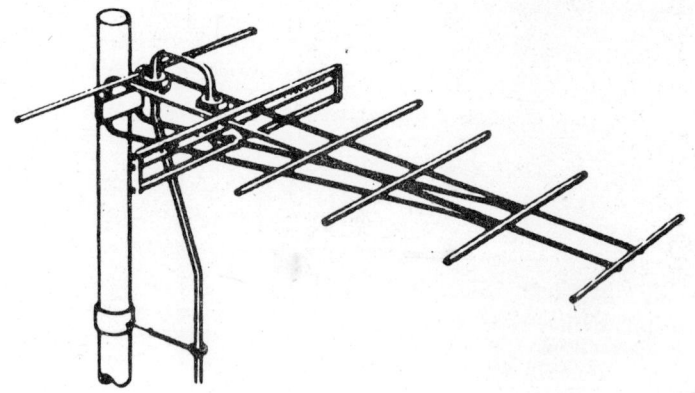

Yagi Antenna

Y-axis.—A line perpendicular to two diametrically opposite parallel faces of a quartz *piezo-electric crystal,* and lying in a plane at right angles to the *Z-axis.* A mechanical axis. *See illustration for X-axis.*
Y-connection.—A *star connection* in a three-phase circuit.
Y-cut.—Descriptive of a *piezo-electric crystal* or quartz plate cut in such manner that a *Y-axis* is perpendicular to its faces. Also called a thirty-degree cut or a face-parallel cut. *See illustration for X-axis.*
yoke.—A deflecting yoke.
Y-signal.—A luminance signal.
Y-waves.—In a *piezo-electric resonator,* waves whose direction is parallel to a *Y-axis.*

Z.

Z. or z.—Symbol for *impedance* in ohms.

Z-axis.—The *optical axis of a piezo-electric crystal*. See illustration for *X-axis*.

Zeppelin antenna.—A *Hertz antenna* system in which the radiator is connected to the transmitter *tank circuit* through an intermediate tuned circuit which in itself is the equivalent of a second Hertz antenna folded at its center. *See illustration.*

zero-angle cut.—An *X-cut* for a quartz crystal.

zero-beat.—The condition in which two frequencies are exactly the same so that they produce no beat frequency when working in

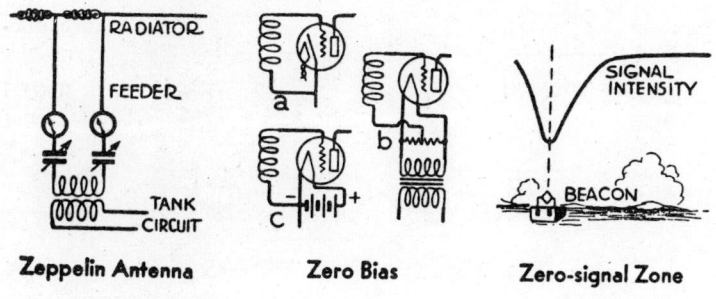

Zeppelin Antenna Zero Bias Zero-signal Zone

the same circuit.

zero beat reception.—Reception by combining with the *carrier wave* a locally generated current of the same frequency, resulting in a current having average changes at the *modulation frequency*. *Homodyne reception*.

zero bias.—A control grid potential which is the same as that of *(a)* the tube's cathode, *(b)* the filament center in an A.C. filament tube, or *(c)* the negative end of the filament in a D.C. filament tube. *See illustration.*

zero level.—A *reference level*.

zero method.—A *null method*.

zero potential or voltage.—The voltage of the earth or of a *ground*.

zero-signal zone.—The region above the antenna of a *radio beacon* in which decrease of signal strength shown by receiving instruments advises a pilot that he has reached the beacon. *See illustration.*

INDUSTRIAL ELECTRONICS

The great majority of words and terms used in all electronic fields are defined in the body of this dictionary. Many definitions in following pages apply especially to industrial and commercial applications of electronic principles and devices.

amplistat.—A *magnetic amplifier.*
analog computer.—A computer whose solution of a problem appears as a graph or plot showing relations between quantities which vary together. For example, were time and temperature the variables, the graph would show temperature to be expected at various intervals of time. Temperature, pressure, speed and other factors may be converted to electric signals which operate the computer. Output from the computer actuates motors or other moving devices in an electro-mechanical graph plotter, or may provide control voltages for a cathode-ray tube which displays the graph. The principal component of the analog computer is an amplifier whose inputs are effectively added, subtracted, multiplied or divided at its output. As an example, to perform addition, several inputs would affect the output, which would represent the sum. A problem is solved by suitably connecting together several elements which multiply, integrate and perform other mathematical processes in the required order.
anemometer.—An instrument for measuring velocity of air motion or for converting velocity to a proportional mechanical or electrical force.
astable multivibrator.—A multivibrator whose operating frequency is not controlled by triggering pulses that start conduction or nonconduction.
automation.—Automatic control, electronic and otherwise of manufacturing or production processes of any kind, including operations on materials, movements between machines or devices in general, assembly, testing and packaging where necessary. Complete automation is not always practicable or economical, it may be combined with manual control or regulation of some steps.

balanced relay.—A *differential relay.*
barretter.—A tube which maintains nearly constant current in a circuit when there are variations of applied voltage.
binary mathematics.—A mathematical system using only two digits as the base, such as 0 and 1, or 1 and 2, instead of the ten digits from 0 to 9 as used in the ordinary decimal system. The binary system is used in computers.
bistable multivibrator.—A multivibrator with which conduc-

INDUSTRIAL ELECTRONICS

tion or nonconduction of each tube or transistor section is started by separate triggering pulses to the two circuits.

Bourbon tube.—A tube formed into part of a circle which expands or contracts with increase or decrease of internal pressure, thus moving the unsupported end of the tube for measurement of or other utilization of changes in pressure or vacuum.

bridge thermometer.—A temperature measuring instrument including a Wheatstone bridge in one of whose arms is an element that changes its resistance with temperature. The null indicator may be graduated for changes of temperature.

capacitance energy storage.—Use of capacitors to accumulate or acquire energy at a relatively slow rate of charge, and to release the total energy at a very high rate of discharge.

capacitance time delay.—A time lapse between opening a circuit and decrease of voltage and current to a minimum, or between closing a circuit and increase to maximum, due to the time-constant effect of respectively discharging or charging a capacitor through a resistance.

carbon film resistor.—*A film resistor.*

chopper.—A device for rapidly interrupting or reversing a steady or slowly changing direct current to produce a rapidly pulsating or alternating current which may be strengthened in any type of a-c amplifier. The amplifier a-c output is rectified or demodulated to recover an amplified reproduction of the direct or slowly changing current originally chopped. Choppers may consist of electromagnetically actuated vibrating reeds with contacts, or of diodes in suitable circuits.

color temperature.—Relation between color of an incandescent substance and its absolute temperature. For example, a certain white may be defined in degrees of absolute temperature at which it appears.

computer.—A machine which, with the help of electronic methods, is capable of solving complete mathematical problems involving addition, subtraction, multiplication, division and other more involved processes. This distinguishes the computer from devices which can handle only one process at a time, such as addition or multiplication. See *analog computer,* also *digital computer.*

constant voltage transformer.—A transformer which maintains a nearly uniform output or secondary voltage when input or primary voltage varies over a certain range.

depletion layer.—A region on each side of a semiconductor junction, as in transistors, from which electrons and holes have been almost entirely withdrawn by action of opposite charges on the junction.

derating.—Reduction of safe power dissipation ability with increase in operating temperature of resistors, capacitors, transistors and other components. Usually expressed as percentages of maximum power capacity permissible for specified numbers of degrees of temperature rise, also for specified increases of power dissipation in watts.

dielectric heating.—*Electrostatic heating.*

differential amplifier.—An amplifier whose output is proportional to the difference between two input signals, which may be of the same or opposite polarity. There may be two separate tubes or transistors with plate or collector outputs combined, or a single tube may have two control grids and one plate.

differential relay.—A relay which operates in accordance with the difference between two currents, or when one current exceeds another by some certain amount.

digital computer.—A computer whose solution of a problem appears as numerical quantities represented by digits, as those from 0 to 9. The problem as presented to the input of the computer also is prepared in the form of digits which represent physical quantities in accordance with some code adapted to the machine. The problem is prepared in the form of successive additions, subtractions, multiplications or divisions which are performed one at a time; this, at least being essentially the method of operation. The solution will appear in numerical relations between variable quantities. For example, with time and temperature the variables, a table would give numbers of degrees of temperature at various times expressed in numbers of seconds, minutes or hours. Accuracy of the digital computer is limited only by the number of digits or "decimal places" to which the solution is carried. Partial solutions of a problem may be examined, and the order or type of remaining mathematical steps altered as appears desirable. The digital computer has a wider field of application than the analog computer.

directional relay.—A relay which alters circuit connections to prevent damage due to reversal of current, voltage, power or phase.

dissipation.—In any current-carrying device the number of watts of power that produces heat which must be removed or dissipated to prevent burnout or other damage. Measured as the product of volts across the device multiplied by amperes of current carried.

drift transistor.—A transistor in whose base element the distribution of impurities is controlled to produce a non-uniform field effect that accelerates electron flow. A type suitable for very-high frequency operation.

electrostatic heating.—Heating of insulating or dielectric materials as a result of dielectric energy losses appearing in the materials when placed in a high-frequency electric or electrostatic field. The energy loss, which changes to heat, is due chiefly to dielectric hysteresis.

electrostriction.—Deformation or strain of certain materials when placed in an electric field. Somewhat similar to magnetostriction caused by a magnetic field.

encapsulated.—Hermetically sealed by a protective covering or coating. Applied to any electrical or electronic component.

equal ampere-turns law.—The number of ampere-turns in primary and secondary windings of a transformer or saturable reactor may be assumed practically equal under all conditions.

feedback.—See *regulation*.

feedback system.—See *servo system*.

field effect transistor.—A two-element transistor in which electron flow between terminals at opposite ends of one element is controlled by a voltage applied to the other element. The control element acts somewhat like the grid of a triode tube, with the current-flow element acting like cathode and plate.

film resistor.—A coating of carbon or of metallic oxides or alloys on a ceramic rod or glass tube, externally protected with a plastic, ceramic or glass covering. Possesses close resistance tolerance. Temperature coefficient of resistance is negative.

flux decay relay.—A time delay relay whose action is slowed by a ring or band of nonmagnetic metal around the coil or part of the core. Fluxes resulting from currents induced in this metal oppose rise and fall of normal core fluxes.

frequency divider.—Commonly two multivibrators with the free running frequency of one lower than that of the other by some definite division or submultiple. Output from the higher frequency unit triggers the other one, but only after capacitor discharge of the other has progressed nearly to the point of reversal of conduction and nonconduction. Thus the lower frequency unit is actuated only once for each certain number of output cycles from the unit of higher frequency, and the higher frequency is effectively divided. Other relaxation oscillators, such as the blocking type, may be similarly used.

frequency multiplier.—Most often a circuit containing one or more tubes, crystal diodes or transistors biased to deliver distorted output waveforms in which are many harmonic frequencies of an applied fundamental frequency. The output circuit is tuned or made resonant at the harmonic frequency that is the desired multiple of the fundamental. Another method employs a full-wave rectifier whose output is two one-way pulses for each alternating cycle or input. The pulses are filtered to form a double-frequency alternating current. Rectifiers may be cascaded for further multiplication. Saturable transformers are used for frequency doublers or triplers.

gate winding.—In a magnetic amplifier or a saturable reactor, a winding which carries output or load current that is controlled or "gated" on or off.

germanium rectifier.—A semiconductor rectifier consisting of N-type and P-type germanium regions. The unit has long operating life when not subjected to excessive temperature.

grid pool tube.—A tube in which the cathode is a pool of mercury. There may be one or more anodes.

gyroscope.—A wheel or disc which, when rapidly rotated, tends to maintain its position in space regardless of external forces applied to the mounting.

hook transistor.—A four-element transistor, usually with the arrangement P-N-P-N. A thin P-type region between two N-type regions effectively traps holes and allows increased electron flow which results in a gain of current between input and output, whereas with triode junction transistors the output collector current does not exceed input emitter current.

high-frequency heating.—Either *induction heating* or *electrostatic heating*.

impulse transformer.—A *peaking transformer*.

induction heating.—Heating or melting of metal due to power dissipated by eddy currents induced in the metal by the high-frequency magnetic field of a coil which surrounds the metal or surrounds a furnace crucible in which is the treated metal.

induction relay.—A time delay relay whose contacts are moved by a shaft carrying an aluminum disc that is rotated by action of a magnetic field produced by coil current and passing through the disc.

inertia relay.—A time delay relay which is slow acting because of weights attached to the moving armature; the weights possess mechanical inertia.

INDUSTRIAL ELECTRONICS

interlocking relay.—A relay having two or more sets of contacts which are caused to open or close together or alternately either by a mechanical connection between armatures or by control of current in one coil by contacts actuated by another coil.

intrinsic barrier transistor.—A four-element transistor in which a thin layer of practically pure material is between two of the three conduction layers such as would be found in an ordinary triode transistor. Useful at very-high and, for some applications, at ultra-high frequencies.

jogging control.—Control for a d-c motor to allow rotation of only a few turns or fraction of a turn between intervals of idleness.

latching relay.—A relay with which current in one electromagnet moves contacts on an armature which is caught and held by a mechanical latch. Release may be by current in a second electromagnet that pulls back on the latch lever. Either latching or release may be by manual means, with the other operation electromagnetic.

limit switch.—An electric switch opened or closed when some part of a moving machine or device comes in contact with the switch arm, button or lever, thus controlling a circuit that limits or otherwise affects the machine action. Usually a *snap action switch*.

locking relay.—A *latching relay*.

magnetic amplifier.—A form of saturable reactor in combination with rectifiers so arranged that small d-c power in one set of windings controls large d-c or a-c powers represented by currents in another set of "gate" windings connected between a power source and a load to be controlled. Thus there is power amplification in the sense that a small power controls a larger one. Where output current or power acts in a large load impedance there is high voltage amplification between input and output.

magnetic brake.—A shoe-type brake applied by spring pressure and released by pull of a solenoid acting against the springs. Also a disc-type brake applied by electromagnets and released by spring pressure.

magnetic particle clutch.—A coupling whose power transfer is controlled by attraction of the rotating members for particles of an iron compound in a magnetic field produced by a coil. A brake may be similarly controlled.

manometer.—An instrument for measurement of small pressures and vacuums as they cause a rise or fall of liquid in one side of a U-shaped tube or other chamber whose other side is open to atmospheric air pressure.

memory system.—In a computer, devices or circuits which store information, such as partial solutions of a problem, until some later time when the information is needed to continue the computations. Ferrite cores magnetized in one polarity or the other may be used.

monostable multivibrator.—A multivibrator in which conduction or nonconduction of only one of its tubes or transistors is started by triggering pulses, the second member operating at a rate determined by its coupling time constants.

multi-turn potentiometer.—A spirally shaped resistance element on which a contact member is moved from end to end by several turns, usually 10 turns, of the shaft. Highly accurate repetitive settings are possible.

Mylar capacitor.—A capacitor with polyester type dielectric, affording high insulation resistance and low dielectric absorption.

notching relay.—A *stepping relay*.

NTC resistor.—An element whose resistance decreases as its temperature rises. A negative temperature coefficient resistor.

peaking transformer.—A transformer with which applied primary voltage causes current which saturates the core before the peak of each primary alternation is reached. This prevents further changes of core flux. The secondary emf (which has been rising) then drops rapidly to zero, to cause alternating brief pulses of secondary current. Used for triggering and other purposes where voltage or current pulses are needed.

phase splitter.—A tube or transistor which, when fed with an alternating voltage, delivers one voltage of the same phase and another of opposite phase. With grid input to a triode the same phase appears at the cathode, and opposite phase at the plate.

photorelay.—A magnetic relay actuated by an amplifier which, in turn, is fed from a phototube or photocell.

pitot tube.—A tube with a small orifice inserted in a pipe or other closed space for measuring velocity of air or other gas in the space.

pickup.—A *transducer*.

plugging control.—Control for a d-c motor which brings it to a very quick stop when power is turned off.

polarized relay.—A relay which operates with coil current in one direction but not in the opposite direction, or with increase in only one direction. Usually consists of an electromagnet and a permanent magnet acting together.

precipitator.—A device for separating and collecting dust or other air-borne particles by means of attraction and repulsion in a strong electrostatic field.

primary element.—A *transducer*.

programming.—In computer operation, a set of instructions fed to the machine for controlling the sequence of operations required for solution of a problem. Magnetic tape, punched cards and other devices are used.

protective tube.—A tube through which current discharges to ground when a connected circuit carries excessive voltage.

pulse transformer.—A *peaking transformer*.

pyrometer.—A device for measuring high temperatures, usually by effects of radiation, of changes in color or by means of a thermocouple.

ratchet relay.—A *stepping relay*.

regulation.—Any method for maintaining uniform performance in an electrical or mechanical system by utilizing the error or deviation to effect a correction. A force derived from error at the output is fed back as a correction to the input. To be distinguished from simple control, with which there is no feedback.

saturable reactor.—A reactor on whose iron core are one or more additional windings in which direct current causes controlled magnetic saturation of the core. The combination forms an adjustable reactance controlled by small d-c power and regulating large a-c power. Principal uses include control of motors, lamps and heating, also voltage or current regulations, operation of solenoids and switching or gating.

saturable transformer.—Two or more a-c windings and one or more additional windings carrying direct current which may be varied to cause magnetic saturation of the core and thereby alter the waveform of output alternating voltage and current.

sensing element.—A *transducer*.

sensitive relay.—A relay which operates with power as small as a few milliwattts or current of a few microamperes in its magnetic coil winding.

sequencing.—Control of the order or sequence in which different operations are performed in manufacturing or processing.

servo motor.—An a-c motor with two sets of stator windings, one of which controls direction and extent of rotation employed for correction in a servo system. Often used in

INDUSTRIAL ELECTRONICS

connection with a magnetic amplifier or saturable transformer feeding the control winding.

servo system.—A method for counteracting errors in operation of a machine or process by applying the effect of any improper changes in output back to the input in such manner as to afford a correction. Output errors may be in speed, position, direction, temperature and such like, which usually are changed by means of a suitable transducer into an electrical or electronic force applied to the input. The input correctional force may be amplified, with the feedback system drawing additional power from an external source. In effect there is a correctional feedback from output to input.

silicon rectifier.—A semiconductor rectifier consisting of N-type and P-type silicon regions. The rectifier withstands higher temperature and greater reverse voltages than germanium rectifiers, but has somewhat greater forward resistance. Used for d-c power supplies operated from a-c lines.

snap action switch.—Electric switch in which current-carrying contacts open and close quickly even though the operating lever or button is moved slowly.

solenoid.—An iron clad electromagnet whose central core is a sliding plunger drawn into the winding by the magnetic effect of coil current. Converts electric power to linear mechanical motion up to several inches where pulling force is desired.

Spacistor.—A transistor with which electron flow in a *depletion layer* is modulated by a signal. The carriers attain high velocity and there is large amplification at high frequencies. There are four terminals, two of which go to the base.

stabilizing.—Prevention of over-control by a servo or other feedback system, an effect which would cause the feedback error force to change the input to an extent causing error in the opposite sense.

stepping relay.—A relay with which each pulse of current in the electromagnet moves a pawl or pivoted arm that engages a toothed wheel or ratchet mounted on a shaft, thus turning the shaft a certain distance. On the shaft may be one or more cams which open and close various pairs of contacts in a certain order or sequence. Several current pulses may be required to complete a control sequence.

storage tube.—An electronic tube capable of receiving computer information and releasing it after a time ranging from a few millionths of a second to minutes. See *memory system*.

strain gage.—A device which utilizes physical deformation or strain to produce an electronic or electrical effect that measures the degree or strain. A type of transducer.

INDUSTRIAL ELECTRONICS

switching.—Rapid reversal or interchange between conductive and nonconductive states of one or more tubes or transistors. Multivibrators are so used, as also are gating tubes of various types. Many transistors are especially suited for switching.

symmetrical transistor.—A type whose collector and emitter elements may be used interchangeably.

synchroscope.—An instrument indicating phase relations between two alternating currents or circuits.

tachometer generator.—An electric generator driven from a shaft whose rotational speed is to be measured. Generator output goes to an indicating meter graduated directly in rotational speeds. There are also electronic forms requiring no mechanical connection to the rotating shaft.

tantalum foil capacitor.—Electrolytic capacitor with plates of tantalum metal foil instead of the more common aluminum foil. Allows reduced size for given capacitance.

tantalum pellet capacitor.—Electrolytic capacitor made with a solid, dry pellet. The dielectric is thin layers of tantalum pentoxide and the electrolyte is managanese dioxide in the form of thousands of grains which are sintered to form the pellet. Properties include long life, safe operation at high temperatures, low leakage, and resistance to moisture, shock and vibration.

thermistor.—A type of NTC resistor whose usual forms are small rods, discs, washers or beads. Used for temperature measurement and control, for time delays, voltage regulation and switching systems operating in accordance with temperature.

Thyrite.—Negative temperature coefficient resistance element made chiefly from silicon carbide.

time delay relay.—Any of many types of relays which open or close their contacts only when a certain time has elapsed after current begins to flow in the electromagnet.

tracer system.—A control for successive steps in operation of a machine. Utilizes limit switches or other on-off elements whose levers or arms follow or trace a cam-like shape formed on a guide or template.

transducer.—Any device which takes energy from one system and supplies energy of the same or any other form to another system. Most often there is a change in the form of energy, as when a microphone changes sound or air-wave energy into variations of electrical energy. Various kinds of transducers deliver an electrical output when the input is a change of pressure, temperature, moisture, position, time, direction, speed and other physical effects. Other transducers change

INDUSTRIAL ELECTRONICS

one mechanical or physical energy to another, as temperature to motion in a thermostat. Still others change one electrical form to another, as varying current to varying reactance in a saturable reactor.

triggering pulse.—Any brief current or voltage whose purpose is to operate or start the operation of some electronic device.

ultrasonic.—Wave motion or vibratory motion at rates or frequencies far above the range of audible sound, often at 200 to 300 kilocycles per second.

Venturi meter.—A device for measuring velocity of a gas flowing through a Venturi tube in which is a tapered internal restriction wherein pressure is proportional to velocity.

Zener diode.—A semiconductor diode of the silicon type with which application of a reverse voltage causes practically no current flow until this voltage reaches a critical "breakdown" value at which there is large current. Often used as a voltage regulator, depending on the fact that voltage drop remains almost constant with large variations of current.